INTERACTIVE MATHEMATICS

## Topic 6
Exponents and Polynomials

## Topic 7
Factoring

## Topic 8
Rational Expressions

## Topic 9
Rational Exponents and Radicals

PERSONAL
ACADEMIC
NOTEBOOK

ACADEMIC
SYSTEMS

Interactive Mathematics



Academic Systems Corporation
444 Castro Street, Suite 1200
Mountain View, CA 94041

800.694.6850

07/98

# TABLE OF CONTENTS

# LESSON 6.1 – EXPONENTS

## OVERVIEW

*Here's what you'll learn in this lesson:*

***Properties of Exponents***

a. *Definition of exponent, power, and base*

b. *Multiplication Property*

c. *Division Property*

d. *Powers raised to powers*

e. *Products raised to powers*

f. *Quotients raised to powers*

g. *The zero exponent*

If a friend agrees to give you one penny today, two pennies tomorrow, four pennies the next day, eight after that, and every day thereafter gives you double the amount he gave you the day before, by the end of a week he will have given you $1.27, and by the end of a month, $10,737,418.23!

While doubling your money every day may not be something you can relate to, it is an example of a situation that can be described using exponents. Another, perhaps more relevant example is the growth of your money in an account earning compound interest.

In this lesson, you will learn some general properties of exponents. These properties will help you later when you learn to simplify expressions involving exponents.

# EXPLAIN

## PROPERTIES OF EXPONENTS

### Summary

You have seen how to use exponential notation as a shortcut for writing multiplication of repeated factors. Now you'll learn some basic properties involving exponents.

### Multiplication Property

There is a rule for multiplying exponential expressions with the same base.

For example, find $3^2 \cdot 3^3$:

$$3^2 \cdot 3^3 = 3 \cdot 3 \cdot 3 \cdot 3 \cdot 3 = 3^{2+3} = 3^5$$

In general:

$$x^m \cdot x^n = x^{m+n}$$

*To multiply two expressions with the same base, add their exponents.*

Here, $m$ and $n$ are positive integers.

This property is called the multiplication property of exponents.

### Division Property

There are also rules for dividing two expressions with the same base.

For example, find $\frac{5^7}{5^4}$:

$$\frac{5^7}{5^4} = \frac{\overset{1}{\cancel{5}} \cdot \overset{1}{\cancel{5}} \cdot \overset{1}{\cancel{5}} \cdot \overset{1}{\cancel{5}} \cdot 5 \cdot 5 \cdot 5}{\underset{1}{\cancel{5}} \cdot \underset{1}{\cancel{5}} \cdot \underset{1}{\cancel{5}} \cdot \underset{1}{\cancel{5}}} = 5^{7-4} = 5^3$$

In general:

$$\frac{x^m}{x^n} = x^{m-n}$$

*To divide two expressions with the same base, subtract their exponents.*

Here, $x \neq 0$, $m$ and $n$ are positive integers, and $m > n$.

You can also divide expressions where the exponent in the denominator is greater than the exponent in the numerator.

For example, find $\frac{5^4}{5^7}$:

$$\frac{5^4}{5^7} = \frac{\overset{1}{\cancel{5}} \cdot \overset{1}{\cancel{5}} \cdot \overset{1}{\cancel{5}} \cdot \overset{1}{\cancel{5}}}{\underset{1}{\cancel{5}} \cdot \underset{1}{\cancel{5}} \cdot \underset{1}{\cancel{5}} \cdot \underset{1}{\cancel{5}} \cdot 5 \cdot 5 \cdot 5} = \frac{1}{5^{7-4}} = \frac{1}{5^3}$$

In general:

$$\frac{x^m}{x^n} = \frac{1}{x^{n-m}}$$

Here, $x \neq 0$, $m$ and $n$ are positive integers, and $m < n$.

These are called the division properties of exponents.

## Power of a Power Property

There is a rule for raising an exponential expression to a power.

For example, find $(2^4)^3$:

$$(2^4)^3 = 2 \cdot 2 \cdot 2 \cdot 2 \quad \cdot \quad 2 \cdot 2 \cdot 2 \cdot 2 \quad \cdot \quad 2 \cdot 2 \cdot 2 \cdot 2 = 2^{4 \cdot 3} = 2^{12}$$

In general:

*When an expression raised to a power is itself raised to a power, multiply the exponents.*

$$(x^m)^n = x^{m \cdot n}$$

This is called the power of a power property of exponents.

## Power of a Product Property

*Here you could have done the multiplication first and then applied the exponent: $(4 \cdot 5)^2 = (20)^2 = 400$. This may seem easier, but when you have an expression that includes variables as well as numbers, the other way is more useful.*

There is a rule for raising a product to a power.

For example, find $(4 \cdot 5)^2$:

$$\begin{aligned} (4 \cdot 5)^2 &= (4 \cdot 5) \cdot (4 \cdot 5) \\ &= 4 \cdot 5 \cdot 4 \cdot 5 \\ &= 4 \cdot 4 \cdot 5 \cdot 5 \\ &= 4^2 \cdot 5^2 \\ &= 16 \cdot 25 \\ &= 400 \end{aligned}$$

In general:

*To raise a product to a power, raise each term to the power.*

$$(xy)^n = x^n \cdot y^n$$

This is called the power of a product property of exponents.

## Power of a Quotient Property

There is a rule for raising a quotient to a power.

For example, find $\left(\frac{2}{5}\right)^3$:

$$\left(\frac{2}{5}\right)^3 = \frac{2}{5} \cdot \frac{2}{5} \cdot \frac{2}{5}$$

$$= \frac{2 \cdot 2 \cdot 2}{5 \cdot 5 \cdot 5}$$

$$= \frac{2^3}{5^3}$$

$$= \frac{8}{125}$$

In general:

$$\left(\frac{x}{y}\right)^n = \frac{x^n}{y^n}$$

Here, $y \neq 0$.

This is called the power of a quotient property of exponents.

*To raise a quotient to a power, raise the numerator and the denominator to the power.*

## Zero Power Property

There is a rule for raising a nonzero quantity to the zero power.

You know that:

$$\frac{x^m}{x^n} = x^{m-n} \text{ when } x \neq 0 \text{ and } m > n$$

and

$$\frac{x^m}{x^n} = \frac{1}{x^{n-m}} \text{ when } x \neq 0 \text{ and } m < n$$

But what happens when $m = n$?

For example, find $\frac{3^2}{3^2}$ in two ways:

1. Calculate the value of the numerator and denominator.
$$\frac{3^2}{3^2} = \frac{3 \cdot 3}{3 \cdot 3} = \frac{9}{9} = 1$$
2. Use the division property of exponents.
$$\frac{3^2}{3^2} = 3^{2-2} = 3^0$$

So, $3^0 = 1$.

In general:

$$x^0 = 1$$

Here, $x \neq 0$.

This is called the zero power property of exponents.

*Any nonzero number raised to the zero power is equal to 1.*

## Sample Problems

1. Find: $b^3 \cdot b^8 \cdot b^6$

☑ a. Add the exponents.
The base stays the same. $= b^{3+8+6}$

$= b^{17}$

2. Find: $\frac{x^9}{x^5}$

☐ a. Subtract the exponent in the denominator from the exponent in the numerator. The base stays the same. = ______

3. Find: $(3^4)^6$

☐ a. Use the power of a power property to multiply the exponents. $= 3^{4 \cdot 6}$
= ______

4. Find: $(b^2 \cdot c)^5$

☐ a. Use the power of a product property to raise each term to the 5th power. = ______

☐ b. Use the power of a power property to multiply the exponents. = ______

5. Find: $\left(\frac{x}{y^3}\right)^2$

☐ a. Use the power of a quotient property to raise the numerator and denominator to the 2nd power. = ______

☐ b. Use the power of a power property to multiply the exponents. = ______

6. Find: $(3x^2y^5)^0$

☐ a. Use the zero power property to evaluate this expression. = ______

***Answers to Sample Problems***

*a.* $x^4$

*a.* $3^{24}$

*a.* $(b^2)^5 \cdot c^5$

*b.* $b^{10} \cdot c^5$ or $b^{10}c^5$

*a.* $\frac{x^2}{y^{3 \cdot 2}}$

*b.* $\frac{x^2}{y^6}$

*a.* 1

# HOMEWORK

## Homework Problems

Circle the homework problems assigned to you by the computer, then complete them below.

### Explain
### Properties of Exponents

Use the appropriate properties of exponents to simplify the expressions in problems 1 through 12. (Keep your answers in exponential form where possible.)

1. Find:

   a. $3^2 \cdot 3^5$ b. $5^2 \cdot 5^5$

   c. $7^2 \cdot 7^5$

2. Find:

   a. $\frac{3^9}{3^5}$ b. $\frac{3^5}{3^9}$

   c. $\frac{3^9}{3^9}$

3. Find:

   a. $(7^3)^2$ b. $(7^2)^3$

4. Find:

   a. $(5 \cdot x)^3$ b. $(3 \cdot y)^2$

   c. $(a^2 \cdot b)^4$

5. Find:

   a. $\left(\frac{x^3 \cdot x^5}{x^4}\right)^2$ b. $\left(\frac{a^{12} \cdot a^6}{a^9 \cdot a^7}\right)^4$

   c. $\left(\frac{b^6 \cdot b^5}{b^3 \cdot b^8}\right)^3$ d. $\frac{2^3 \cdot x^5}{2^5 \cdot x^2}$

6. Find:

   a. $(a^2 \cdot a^3)^2 + (a^2 \cdot a^3)^2$

   b. $\frac{y^4 \cdot 3y^2}{y^8}$

   c. $x^4 \cdot x^9 \cdot x \cdot y^5 \cdot y^{11}$

7. Find:

   a. $(b^3)^2 \cdot (b^4)^3$

   b. $\frac{y^6}{y^{17}} \cdot (y^5)^2 \cdot (y^3)^4$

   c. $\frac{a^4 \cdot b^6}{a^{11} \cdot b^3}$

8. Find:

   a. $\frac{(xy)^4}{y^9 \cdot x^7}$ b. $\frac{(3b)^6}{(3b^2)^4}$

9. As animals grow, they get taller faster than they get stronger. In general, this proportion of increase in height to increase in strength can be written as $\frac{x^2}{x^3}$. Simplify this fraction.

10. An animal is proportionally stronger the smaller it is. If a person is 200 times as tall as an ant, figure out how much stronger a person is, pound for pound, by simplifying the expression $\frac{200^2}{200^3}$.

11. Find:

    a. $\left(\frac{4xy^2z}{5x^2yz^3}\right)^0$ b. $\frac{y^7 \cdot y}{y^9 \cdot y^2}$

    c. $\left(\frac{b^3 \cdot b^5}{b^6 \cdot b^3}\right)^4$ d. $-2x^0 + 4y^0$

12. Find:

    a. $\left(\frac{(x^3 \cdot x^4)^2}{x^7}\right)^5$ b. $\frac{(4a^2)^0 - 3b^0}{2}$

    c. $\left(\frac{(3x \cdot 3x^2)^2}{3^{11} \cdot x^7}\right)^3$ d. $\left(\frac{b^8}{(b^2 \cdot b^7)^3}\right)^4$

# APPLY

## Practice Problems

Here are some additional practice problems for you to try.

### Properties of Exponents

1. Find: $7^5 \cdot 7^3$. Leave your answer in exponential notation.
2. Find: $6^3 \cdot 6^4$. Leave your answer in exponential notation.
3. Find: $b^{12} \cdot b^3$
4. Find: $c^9 \cdot c^4$
5. Find: $a^6 \cdot a^5$
6. Find: $5^7 \div 5^3$. Leave your answer in exponential notation.
7. Find: $9^{10} \div 9^4$. Leave your answer in exponential notation.
8. Find: $\frac{m^{10}}{m^4}$
9. Find: $\frac{n^{20}}{n^{15}}$
10. Find: $\frac{b^{12}}{b^5}$
11. Find: $(5^3)^4$. Leave your answer in exponential notation.
12. Find: $(8^2)^5$. Leave your answer in exponential notation.
13. Find: $(13^5)^6$. Leave your answer in exponential notation.
14. Find: $(y^8)^3$
15. Find: $(z^{12})^4$
16. Find: $(x^9)^4$
17. Find: $(3 \cdot a)^4$
18. Find: $(4 \cdot b)^2$
19. Find: $(2 \cdot y)^3$
20. Find: $\frac{a^6b^5}{a^8b^2}$
21. Find: $\frac{m^7n^4}{m^3n^{10}}$
22. Find: $\frac{x^3y^7z^{12}}{xy^8z^5}$
23. Find: $5^0$
24. Find: $348^0$
25. Find: $x^0$
26. Find: $5^1 + (4z)^0$
27. Find: $a^0 - (xyz)^0 + 3^1$
28. Find: $2^1 - (3x)^0 + y^0$

# EVALUATE

## Practice Test

Take this practice test to be sure that you are prepared for the final quiz in Evaluate.

1. Rewrite each expression below. Keep your answer in exponential form where possible.

   a. $11 \cdot 11 \cdot 11 \cdot 11$

   b. $3 \cdot 3 \cdot y \cdot y \cdot y \cdot y \cdot y$

   c. $5^{12} \cdot 5^{8} \cdot 5^{23}$

   d. $x^{7} \cdot y \cdot y^{19} \cdot x^{14} \cdot y^{6}$

   e. $7^{8} \cdot b^{5} \cdot b^{8} \cdot 7^{10} \cdot b$

2. Rewrite each expression below in simplest form using exponents.

   a. $\frac{2 \cdot 2 \cdot 2 \cdot 2 \cdot 2 \cdot 2}{2 \cdot 2 \cdot 2}$

   b. $\frac{b^{20}}{b^{14}}$

   c. $\frac{3^{12} \cdot x^{7}}{3^{9} \cdot x^{16}}$

   d. $\frac{y^{17}}{y^{14} \cdot y^{3} \cdot y^{4}}$

3. Circle the expressions below that simplify to $\frac{x^3}{y^5}$.

   $\frac{x^6y^2}{x^3y^7}$ $\qquad$ $\frac{y^{11}x^5}{y^2x^4}$

   $\frac{xy^9}{x^6y^4}$ $\qquad$ $\frac{x^7y}{x^4y^6}$

4. Circle the expressions below that simplify to $5y$.

   $(31x^{8})^{0} \cdot 5y$

   $-(-5y)^{0}$

   $\frac{5y^2}{y}$

   $\frac{(5y)^2}{5y}$

   $\frac{5 \cdot 5 \cdot 5 \cdot y \cdot y \cdot y \cdot y}{5 \cdot 5 \cdot y \cdot y}$

5. Simplify each expression below.

   a. $(b^{4} \cdot b^{2})^{8}$

   b. $(3^{5} \cdot a^{6})^{2}$

   c. $(2^{9} \cdot x^{4} \cdot y^{6})^{11}$

6. Simplify each expression below.

   a. $\left(\frac{5y^{10}}{3x^{8}}\right)^{4}$

   b. $\left(\frac{7a^{3}b^{4}}{5a^{2}}\right)^{6}$

7. Calculate the value of each expression below.

   a. $(4x)^{0} - 2y^{0}$

   b. $(5xy^{2} \cdot 4x^{3})^{0}$

   c. $-2x^{0} - y^{0}$

   d. $\frac{(4x)^0}{2} + \frac{3x^0}{2} + \frac{-2x^0}{2}$

8. Rewrite each expression below using a single exponent.

   a. $\left(\frac{a^{4} \cdot a^{5}}{a \cdot a^{3}}\right)^{7}$

   b. $\left(\frac{a \cdot a^{3}}{a^{4} \cdot a^{5}}\right)^{7}$

# ANSWERS

## Homework

**1a.** $3^7$ **b.** $5^7$ **c.** $7^7$ **3a.** $7^6$ **b.** $7^6$

**5a.** $x^8$ **b.** $a^8$ **c.** 1 **d.** $\frac{x^3}{4}$ **7a.** $b^{18}$ **b.** $y^{11}$ **c.** $\frac{b^3}{a^7}$

**9.** $\frac{1}{x}$ **11a.** 1 **b.** $\frac{1}{y^3}$ **c.** $\frac{1}{b^4}$ **d.** 2

## Practice Problems

**1.** $7^8$ **3.** $b^{15}$ **5.** $a^{11}$ **7.** $9^6$ **9.** $n^5$

**11.** $5^{12}$ **13.** $13^{30}$ **15.** $z^{48}$ **17.** $81a^4$

**19.** $8y^3$ **21.** $\frac{m^4}{n^6}$ **23.** 1 **25.** 1 **27.** 3

## Practice Test

**1a.** $11^4$ **b.** $3^2y^5$ **c.** $5^{43}$ **d.** $x^{21}y^{26}$ **e.** $7^{18}b^{14}$

**2a.** $2^3$ **b.** $b^6$ **c.** $\frac{3^3}{x^9}$ **d.** $\frac{1}{y^4}$

**3.** $\frac{x^6y^2}{x^3y^7}$ and $\frac{x^7y}{x^4y^6}$

**4.** $(31x^8)^0 \cdot 5y$, $\frac{5y^2}{y}$ and $\frac{(5y)^2}{5y}$

**5a.** $b^{48}$ **b.** $3^{10}a^{12}$ **c.** $2^{99}x^{44}y^{66}$

**6a.** $\frac{5^4y^{40}}{3^4x^{32}}$ **b.** $\frac{7^6a^6b^{24}}{5^6}$

**7.a.** $-1$ **b.** 1 **c.** $-3$ **d.** 1

**8a.** $a^{35}$ **b.** $\frac{1}{a^{35}}$

# LESSON 6.2 – POLYNOMIAL OPERATIONS I

# OVERVIEW

***Here's what you'll learn in this lesson:***

***Adding and Subtracting***

*a. Definition of polynomial, term, and coefficient*

*b. Evaluating a polynomial*

*c. The degree of a term and a polynomial*

*d. Writing the terms of a polynomial in descending order*

*e. Definition of a monomial, binomial, and trinomial*

*f. Recognizing like or similar terms*

*g. Combining like or similar terms*

*h. Polynomial addition*

*i. Polynomial subtraction*

***Multiplying and Dividing***

*a. Multiplying a monomial by a monomial*

*b. Multiplying a monomial by a polynomial*

*c. Dividing a monomial by a monomial*

*d. Dividing a polynomial by a monomial*

Every day, people use algebra to find unknown quantities. For example, you may be interested in figuring out how long it would take you to drive across the country. Or, you may want to know why your checkbook doesn't balance.

To find these unknown quantities, you need to be able to add, subtract, multiply, and divide polynomials. That's what you will learn in this lesson.

# EXPLAIN

## ADDING AND SUBTRACTING

### Summary

#### Identifying Polynomials

A polynomial is a special kind of algebraic expression which may have one or more variables and one or more terms.

For a polynomial in one variable, $x$, each term has the form $ax^r$, where the coefficient, $a$, is any real number, and the exponent, $r$, is a nonnegative integer.

*Remember, when $x \neq 0$:*

$x^0 = 1$

$x^1 = x$

For example:

$$x^3 + 7x^2 - 4x + 2$$

| $x^3$ | $7x^2$ | $-4x$ | $2$ |
|---|---|---|---|
| $a = 1$ | $a = 7$ | $a = -4$ | $a = 2$ |
| $r = 3$ | $r = 2$ | $r = 1$ | $r = 0$ |

For a polynomial in two variables, $x$ and $y$, each term has the form $ax^ry^s$, where $a$ is any real number, and $r$ and $s$ are nonnegative integers.

For example:

$$8x^3y^4 + 3xy^2 - 4$$

| $8x^3y^4$ | $3xy^2$ | $-4$ |
|---|---|---|
| $a = 8$ | $a = 3$ | $a = -4$ |
| $r = 3$ | $r = 1$ | $r = 0$ |
| $s = 4$ | $s = 2$ | $s = 0$ |

Polynomials with one, two, or three terms have special names:

A monomial has one term: $\frac{1}{4}t^5u^3v^2$

A binomial has two terms: $4mpd^2 + 2m^3d$

A trinomial has three terms: $-87k - 13k^2 + \sqrt{13}k^3$

An algebraic expression is not a polynomial if any of its terms cannot be written in the form $ax^r$.

For example, these algebraic expressions are **not** polynomials:

$\frac{2}{3w} + 7w^2$ $\qquad \sqrt{5t^2r} + 2t^3$ $\qquad 6x^2 - 2\sqrt{y}$

## The Degree of a Polynomial

The degree of a term of a polynomial is the sum of the exponents of the variables in that term. The degree of a polynomial is the degree of the term with the highest degree.

For example, to find the degree of the polynomial $8x^3y^4 + 3xy^2 - 3$; find the degree of each term:

$$8x^3y^4 + 3xy^2 - 3 = \underset{7}{8x^3y^4} + \underset{3}{3x^1y^2} - \underset{0}{3x^0y^0}$$

The degree of the polynomial $8x^3y^4 + 3xy^2 - 3$ is the degree of the term with the highest degree, 7.

## Evaluating Polynomials

Sometimes the variables in a polynomial are assigned specific numerical values. In these cases you can evaluate the polynomial by replacing the variables with the numbers.

To evaluate a polynomial:

1. Replace each variable with its assigned value.

2. Calculate the value of the polynomial.

For example, to evaluate the polynomial $6b^2 - 4bc + c^2$ when $b = 2$ and $c = 3$:

| | $6b^2 - 4bc + c^2$ |
|---|---|
| 1. Replace $b$ with 2 and $c$ with 3. | $= 6(2)^2 - 4(2)(3) + (3)^2$ |
| 2. Calculate. | $= 24 - 24 + 9$ |
| | $= 9$ |

So, when $b = 2$ and $c = 3$, $6b^2 - 4bc + c^2 = 9$.

## Adding Polynomials

To add polynomials, combine like terms - terms that have the same variables with the same exponents.

Here is an example of two like terms:

$3x^2y$ and $-2x^2y$

Here is an example of two terms that are not like terms:

$3x^2y$ and $3xy^2$

To add polynomials:

1. Remove the parentheses.
2. Write like terms next to each other.
3. Combine like terms.

For example, to find: $(3z^3 + 2zy^2 - 6y^3) + (15z^2 - 5zy^2 + 4z^2)$

1. Remove the parentheses. $= 3z^3 + 2zy^2 - 6y^3 + 15z^2 - 5zy^2 + 4z^2$
2. Write like terms next to each other. $= 3z^3 + 2zy^2 - 5zy^2 - 6y^3 + 15z^2 + 4z^2$
3. Combine like terms. $= 3z^3 - 3zy^2 - 6y^3 + 19z^2$

## Subtracting Polynomials

To subtract one polynomial from another, add the opposite of the polynomial being subtracted.

To subtract polynomials:

1. Multiply the polynomial being subtracted by –1.
2. Distribute the –1.
3. Simplify.
4. Write like terms next to each other.
5. Combine like terms.

For example, to find: $(6y^3 - 3z^3 + 2zy^2) - (15z^2 - 5zy^2 + 4y^3)$

1. Multiply the second polynomial by –1. $= (6y^3 - 3z^3 + 2zy^2) + (-1)(15z^2 - 5zy^2 + 4y^3)$
2. Distribute the –1. $= (6y^3 - 3z^3 + 2zy^2) + (-1)15z^2 - (-1)5zy^2 + (-1)4y^3$
3. Simplify. $= 6y^3 - 3z^3 + 2zy^2 - 15z^2 + 5zy^2 - 4y^3$
4. Write like terms next to each other. $= 6y^3 - 4y^3 - 3z^3 + 2zy^2 + 5zy^2 - 15z^2$
5. Combine like terms. $= 2y^3 - 3z^3 + 7zy^2 - 15z^2$

*To find the opposite of a polynomial, multiply each term by –1.*

*When you add the opposite, the result is the same as changing the sign of each term in the polynomial being subtracted.*

***Answers to Sample Problems***

*b. −2, −2*

*c. 250, 60, −6*

*321*

*b. $3x^2$, $5xy^2$, 3*

*c. $4x^2$, $7xy^2$, 5*

*b. $3x^2$, $-5xy^2$, 3*

*c. $x^2 - 2xy^2 + 2 - 3x^2 + 5xy^2 - 3$*

*d. $x^2 - 3x^2 - 2xy^2 + 5xy^2 + 2 - 3$ (in any order that like terms are next to each other)*

*e. $-2x^2 + 3xy^2 - 1$ (in any order)*

# Sample Problems

1. Evaluate the polynomial $2r^3 + 3s^2r - 3s + 5$ when $r = 5$ and $s = -2$.

   Evaluate: $2r^3 + 3s^2r - 3s + 5$

   ☑ a. Replace $r$ with 5. $= 2(5)^3 + 3s^2(5) - 3s + 5$

   ☐ b. Replace $s$ with −2. $= 2(5)^3 + 3(\_\_\_)^2(5) - 3(\_\_\_) + 5$

   ☐ c. Calculate. $= \_\_\_ + \_\_\_ - \_\_\_ + 5$

   $= \_\_\_$

2. Find: $(x^2 - 2xy^2 + 2) + (3x^2 - 5xy^2 + 3)$

   ☑ a. Remove the parentheses. $= x^2 - 2xy^2 + 2 + 3x^2 - 5xy^2 + 3$

   ☐ b. Write like terms next to each other. $= x^2 + \_\_\_ - 2xy^2 - \_\_\_ + 2 + \_\_\_$

   ☐ c. Combine like terms. $= \_\_\_ - \_\_\_ + \_\_\_$

3. Find: $(x^2 - 2xy^2 + 2) - (3x^2 - 5xy^2 + 3)$

   ☑ a. Multiply each term in the second polynomial by −1. $= (x^2 - 2xy^2 + 2) + (-1)(3x^2 - 5xy^2 + 3)$

   ☐ b. Distribute the −1.
   $= (x^2 - 2xy^2 + 2) + (-1)(\_\_\_) + (-1)(\_\_\_) + (-1)(\_\_\_)$

   ☐ c. Simplify. $= \_\_\_\_\_\_\_\_$

   ☐ d. Write like terms next to each other.
   $= \_\_\_\_\_\_\_\_$

   ☐ e. Combine like terms.
   $= \_\_\_\_\_\_\_\_$

# MULTIPLYING AND DIVIDING

## Summary

### Multiplying Monomials

Monomials are easy to multiply because they each have only one term.

To multiply monomials:

1. Rearrange the factors so that the constants are next to each other and the factors with the same base are next to each other.
2. Multiply.

For example, to find: $-3p^3q^6r^2s \cdot 2p^4q^4s^5$

1. Rearrange the factors. $= -3 \cdot 2 \cdot p^3 \cdot p^4 \cdot q^6 \cdot q^4 \cdot r^2 \cdot s \cdot s^5$
2. Multiply. $= -6 \cdot p^{3+4} \cdot q^{6+4} \cdot r^2 \; s^{1+5}$

$= -6 \cdot p^7 \cdot q^{10} \cdot r^2 \cdot s^6$

$= -6p^7q^{10}r^2s^6$

### Multiplying a Monomial by a Polynomial with More Than One Term

When multiplying a monomial by a polynomial with more than one term, you need to multiply every term in the polynomial by the monomial.

To multiply a monomial by a polynomial with more than one term:

1. Distribute the monomial to each term in the other polynomial.
2. Simplify.

For example, to find: $5x^4(3x^2y^2 + 2xy^2 - x^2y)$

1. Distribute the monomial. $= 5x^4(3x^2y^2) + 5x^4(2xy^2) - 5x^4(x^2y)$

$= 5 \cdot 3 \cdot x^4 \cdot x^2 \cdot y^2 + 5 \cdot 2 \cdot x^4 \cdot x \cdot y^2 - 5 \cdot x^4 \cdot x^2 \cdot y$

2. Simplify. $= 15 \cdot x^{4+2} \cdot y^2 + 10 \cdot x^{4+1} \cdot y^2 - 5 \cdot x^{4+2} \cdot y$

$= 15x^6y^2 + 10x^5y^2 - 5x^6y$

*In general, to multiply a monomial by a polynomial with more than one term:*

*$a(b + c + d) = a \cdot b + a \cdot c + a \cdot d$*

*Remember:*

*When* $r > s$, $x^r \div x^s = \frac{x^r}{x^s} = x^{r-s}$.

*When* $r < s$, $x^r \div x^s = \frac{x^r}{x^s} = \frac{1}{x^{s-r}}$.

*The y is on the bottom since the exponent of y in the denominator is bigger than the exponent of y in the numerator.*

## Dividing Monomials

To divide monomials:

1. Write the division as a fraction.
2. Cancel common numerical factors.
3. Divide factors with the same base by subtracting exponents.

For example, to find $4x^3y^4z^5 \div 10x^2y^6z^2$:

1. Write the division as a fraction. $= \frac{4x^3y^4z^5}{10x^2y^6z^2}$
2. Cancel common numerical factors. $= \frac{\overset{1}{\cancel{2}} \cdot 2x^3y^4z^5}{\underset{1}{\cancel{2}} \cdot 5x^2y^6z^2}$
3. Divide factors with the same base. $= \frac{2x^{3-2}z^{5-2}}{5y^{6-4}}$

$= \frac{2x^1z^3}{5y^2}$

## Dividing a Polynomial with More Than One Term by a Monomial

When you divide a polynomial with more than one term by a monomial, you must divide each term of the polynomial by the monomial.

Use this rule: $\frac{a+b}{c} = \frac{a}{c} + \frac{b}{c}$.

To divide a polynomial with more than one term by a monomial:

1. Write the division as a fraction.
2. Rewrite the fraction using the rule $\frac{a+b}{c} = \frac{a}{c} + \frac{b}{c}$.
3. Perform the division on each of the resulting terms.

For example, to find $(15x^6y^2 + 10x^5y^3) \div 5x^4y$:

1. Write the division as a fraction. $= \frac{15x^6y^2 + 10x^5y^3}{5x^4y}$
2. Rewrite the fraction using the rule $\frac{a+b}{c} = \frac{a}{c} + \frac{b}{c}$. $= \frac{15x^6y^2}{5x^4y} + \frac{10x^5y^3}{5x^4y}$
3. Divide each of the resulting terms. $= \frac{3 \cdot \overset{1}{\cancel{5}}x^{6-4}y^{2-1}}{\underset{1}{\cancel{5}}} + \frac{2 \cdot \overset{1}{\cancel{5}}x^{5-4}y^{3-1}}{\underset{1}{\cancel{5}}}$

$= 3x^2y + 2xy^2$

## Sample Problems

1. Find: $3wx^3y^6 \cdot 7wx^2y^5z^2$

☑ a. Rearrange the factors so the constants are next to each other and factors with the same base are next to each other.

$= 3 \cdot 7 \cdot w \cdot w \cdot x^3 \cdot x^2 \cdot y^6 \cdot y^5 \cdot z^2$

☐ b. Multiply factors with the same base by adding the exponents.

$= 21 \cdot w^{\_\_} \cdot x^{\_\_} \cdot y^{\_\_} \cdot z^{\_\_}$

2. Find: $pr^2s(p^2r + pr^3s^6 - 2)$

☐ a. Distribute the monomial.

$= \_\_\_\_\_\_ \cdot p^2r + \_\_\_\_\_\_ \cdot pr^3s^6 - \_\_\_\_\_\_ \cdot 2$

☐ b. Multiply each of the resulting terms.

$= p^{1+2}r^{2+1}s + p^{1+1}r^{2+3}s^{1+6} - 2pr^2s$

$= \_\_\_\_\_\_\_\_ + \_\_\_\_\_\_\_\_ - 2pr^2s$

3. Find: $18m^6n^5p^3r \div 10m^3n^7pr$

☑ a. Write the division as a fraction.

$= \frac{18m^6n^5p^3r}{10m^3n^7pr}$

☑ b. Cancel common numerical factors.

$= \frac{\overset{1}{\cancel{2}} \cdot 9m^6n^5p^3r}{\underset{1}{\cancel{2}} \cdot 5m^3n^7pr}$

☐ c. Divide factors with the same base by subtracting exponents.

$= \_\_\_\_\_\_\_\_$

4. Find: $(6w^5x^3 + 4w^3x^2y) \div 2w^2xy$

☑ a. Write the division as a fraction.

$= \frac{6w^5x^3 + 4w^3x^2y}{2w^2xy}$

☐ b. Rewrite the fraction using the rule $\frac{a+b}{c} = \frac{a}{c} + \frac{b}{c}$.

$= \frac{6w^5x^3}{2w^2xy} + \_\_\_\_\_\_$

☐ c. Divide each term.

$= \frac{3w^3x^2}{y} + \_\_\_\_\_\_$

***Answers to Sample Problems***

*1. b.* 2, 5, 11, 2

*2. a.* $pr^2s, pr^2s, pr^2s$

*2. b.* $p^3r^3s, p^2r^5s^7$

*3. c.* $\frac{9m^3p^2}{5n^2}$

*4. b.* $\frac{4w^3x^2y}{2w^2xy}$

*4. c.* $2wx$

# HOMEWORK

## Homework Problems

Circle the homework problems assigned to you by the computer, then complete them below.

### Explain

### Adding and Subtracting

1. Circle the algebraic expression that is a polynomial.

   $3\frac{1}{4}y^3 + \sqrt{3y^2 - 5}$

   $3\frac{1}{4}y^3 + 3y^2 - \sqrt{5}$

   $\frac{1}{4y^3} + 3y^2 - 5$

2. Write m beside the monomial, b beside the binomial, and t beside the trinomial.

   ____ $34x + x^2 + z$

   ____ $wxy^3z^2$

   ____ $pn^2 - 13n^3$

3. Given the polynomial $3y - 2y^3 - 4y^5 + 2$:

   a. write the terms in descending order.

   b. find the degree of each term.

   c. find the degree of the polynomial.

4. Find: $(-3w - 12w^3 + 2) + (15w - 2w^3 + 4w^5 - 3)$

5. Find: $(2v^3 + 6v^2 + 2) - (5v + v^3 + 4v^7 - 3)$

6. Evaluate $\frac{1}{4}xy + 3y^2 - 5x^3$ when $x = 2$ and $y = 4$.

7. Find:
   $(-s^2t + s^3t^3 + 4st^2 - 27) + (3st^2 + 2st - 8s^3t^3 - 13t + 36)$

8. Find: $(12x^3y + 9x^2y^2 + 6xy - y + 7) - (7xy - x + y - 11x^3y + 3x^2y^2 - 4)$

9. Angelina works at a pet store. Today, she is cleaning three fish tanks. These polynomials describe the volumes of the tanks:

   Tank 1: $xy^2$

   Tank 2: $x^2y - 2y^3 + 4xy^2 + 3$

   Tank 3: $x^2y + 5xy^2 + 6y^3$

   Write a polynomial that describes the total volume of the three tanks.

   Hint: Add the polynomials.

   volume = ________

10. Angelina has three fish tanks to clean. These polynomials describe their volumes.

    Tank 1: $xy^2$

    Tank 2: $x^2y - 2y^3 + 4xy^2 + 3$

    Tank 3: $x^2y + 5xy^2 + 6y^3$

    What is the total volume of the fish tanks if $x = 3$ feet and $y = 1.5$ feet?

    volume = ________ cubic feet

11. Find:
$(w^2yz + 3w^3 - 2wyz^2 + 4wyz) - (4wy^2z - 3w^2yz + 2wyz^2) + (2wyz + 3)$

12. Find:
$(tu^2v - 4t^2u^2v + 9t^3uv + 3tv) + (3t^2u^2 + 2tv - t^3) - (4t^2u^2v + 3tv + 2tu^2v) - (6t^3uv + 2tv)$

## Multiplying and Dividing

13. Find: $xyz \cdot x^2y^2z^2$

14. Find: $3p^2r \cdot 2p^3qr$

15. Find: $-6t^3u^2v^{11} \cdot \frac{1}{2}tu^2v^4$

16. Find: $3y(2x^3 + 3x^2y)$

17. Find: $5p^2r^3(2pr + p^2r^2)$

18. Find: $t^3uv^4(2tu - 3uv + 4tv + 5)$

19. Write $12w^7x^3y^2z^6 \div 4w^2x^2y^3z^6$ as a fraction and simplify.

20. Write $(36x^3y^3 + 15x^2y^5) \div 9x^2y$ as a fraction and simplify.

21. Find: $15a^7b^4d^2 \div 10a^4b^9c^3d$

22. Tony is an algebra student. This is how he answered a question on a test:
$(2t^8u^3 - 4t^4u^9 + 6t^{12}u^6) \div 2t^4u^3 = t^2u - 2tu^3 + 3t^3u^2$

Is his answer right or wrong? Why? Circle the most appropriate response.

The answer is right.

The answer is wrong. Tony divided the exponents rather than adding them. The correct answer is $t^{12}u^6 - t^8u^{12} + t^{16}u^9$.

The answer is wrong. The terms need to be ordered by degree. The correct answer is $3t^3u^2 + t^2u - 2tu^3$.

The answer is wrong. Tony divided the exponents rather than subtracting them. The correct answer is $t^4 - 2u^6 + 3t^8u^3$.

The answer is wrong. Tony shouldn't have canceled the numerical coefficients. The correct answer is $2t^2u - 4tu^3 + 6t^3u^2$.

23. Find: $(16x^2y^4 + 20x^3y^5) \div 12xy^2$

24. Find: $(20t^5u^{11} + 5t^3u^5 + 30tu^6v^5) \div 10t^4u^5$

# APPLY

## Practice Problems

Here are some additional practice problems for you to try.

### Adding and Subtracting

1. Circle the algebraic expressions below that are polynomials.

   $2xy + 5xz$

   $\frac{2}{3x} + 6x$

   $9y^2 + 13yz - 8z^2$

   $\sqrt{24x^5}$

   $\frac{15a^3}{5a^8}$

2. Circle the algebraic expressions below that are polynomials.

   $8xy + \frac{3}{y}$

   $\sqrt{17x^3}$

   $3w - 7wz - 1$

   $7x^2 - 13x + 8y^2$

   $\frac{12x^2}{3x^3}$

3. Identify each polynomial below as a monomial, a binomial, or a trinomial.

   a. $17x + 24z$

   b. $13ab^2 - 5$

   c . $m - n + 10$

   d. $42a^2b^4c$

   e. $73 + 65x - 21y$

4. Identify each polynomial below as a monomial, a binomial, or a trinomial.

   a. $25 - 6xyz - 4x$

   b. $2xyz^3$

   c. $x + y - 1$

   d. $36 - 3xyz$

   e. $32x^2y$

5. Find the degree of the polynomial $8a^3b^5 - 11a^2b^3 + 7b^6$.

6. Find the degree of the polynomial $12m^4n^7 - 16m^{12}$.

7. Find the degree of the polynomial $7x^3y^2z + 3x^2y^3z^4 + 6z^7$.

8. Evaluate $2x^2 - 8x + 11$ when $x = -1$.

9. Evaluate $x^3 + 3x^2 - x + 1$ when $x = -2$.

10. Evaluate $2x^2 - 5x + 8$ when $x = 3$.

11. Evaluate $x^2y + xy^2$ when $x = 2$ and $y = -3$.

12. Evaluate $5mn + 4mn^2 + 8m - n$ when $m = 4$ and $n = -2$.

13. Evaluate $3uv - 6u^2v + 2u - v + 4$ when $u = 2$ and $v = -4$.

14. Find: $(3x^2 + 7x) + (x^2 - 5)$

15. Find: $(5x^2 + 4x - 8) + (x^2 + 7x)$

16. Find: $(6a^2 + 8a - 10) + (-3a^2 - 2a + 7)$

17. Find: $(12m^2n^3 + 7m^2n^2 - 14mn) + (3m^2n^3 - 5m^2n^2 + 7mn)$

18. Find: $(10x^4y^3 - 9x^2y^3 + 6xy^2 - x) + (-8x^4y^3 + 14x^2y^3 + 3xy^2 + x)$

19. Find: $(13a^3b^2 + 6a^2b - 5ab^3 + b) + (2a^3b^2 - 2a^2b + 4ab^3 - b)$

20. Find: $(11u^5v^4w^3 + 6u^3v^2w) + (6u^5v^4w^3 - 11u^3v^2w)$

21. Find: $(7xy^2z^3 - 19x^2yz^2 + 26x^3y^3z) + (13xy^2z^3 - 11x^2yz^2 - 16x^3y^3z)$

22. Find: $(9a^4b^2c - 3a^2b^3c + 5abc) + (2abc - 6a^4b^2c - 2) + (3a^2b^3c + 5)$

23. Find: $(5x^3 + 7x) - (x^3 + 8)$

24. Find: $(9a^2 + 7ab + 14b) - (3a^2 - 7b)$

25. Find: $(2y^2 + 6xy + 3y) - (y^2 - y)$

26. Find: $(8x^3 + 9x^2 + 17) - (5x^3 - 3x^2 + 15)$

27. Find: $(9a^5b^3 + 8a^4b - 6b) - (-2a^5b^3 + 12a^4b + 3b)$

28. Find: $(7x^4y^2 - 3x^2y + 5x) - (9x^4y^2 + 3x^2y - 2x)$

## Multiplying and Dividing

29. Find: $3y^4 \cdot 5y$

30. Find: $5x^3 \cdot 2x$

31. Find: $-5a^5 \cdot 9a^4$

32. Find: $-3x^3 \cdot 12x^4$

33. Find: $4x^3y^5 \cdot 7xy^3$

34. Find: $-7a^5b^6c^3 \cdot 8ab^3c$

35. Find: $-3w^2x^3y^2z \cdot 2x^2yz^2$

36. Find: $4y^3(3y^2 + 5y - 10)$

37. Find: $-2a^3b^2(3a^4b^5 - 5ab^3 + 6a)$

38. Find: $2xy^3(2x^6 - 5x^4 + y^2)$

39. Find: $5a^2b^2(4a^2 + 2a^2b - 7ab^2 - 3b)$

40. Find: $-4mn^3(-3m^2n + 12mn^2 - 6m + 7n^2)$

41. Find: $4x^3y^3(3x^3 - 7xy^2 + 2xy - y)$

42. Find: $\frac{9x^3y}{3x^2}$

43. Find: $\frac{20a^5b^6}{4a^3b}$

44. Find: $\frac{12x^4y^6}{3x^2y}$

45. Find: $\frac{32a^7b^9c}{12a^5b^6c^2}$

46. Find: $\frac{15m^6n^{10}}{10n^4p^3}$

47. Find: $\frac{24x^6y^2z^7}{16wx^3z^2}$

48. Find: $\frac{27a^4b^3c^{12}d}{15ac^7d^3}$

49. Find: $\frac{42mn^6p^3q^4}{28m^2nq^5}$

50. Find: $\frac{36w^2x^3y^7z}{21w^5y^2z^2}$

51. Find: $\frac{32a^3 + 24a^5}{8a^2}$

52. Find: $\frac{21m^2 + 18mn^3}{3mn}$

53. Find: $\frac{14x + 8x^4y^2}{2xy}$

54. Find: $\frac{24a^2b^2c^3 - 4ab^4c^5}{16abc^3}$

55. Find: $\frac{32x^2y^3z^4 - 8x^5yz^7}{16x^3y^3z^4}$

56. Find: $\frac{32r^4st^2 - 3r^2st^5}{12r^3s^2t}$

# EVALUATE

## Practice Test

Take this practice test to be sure that you are prepared for the final quiz in Evaluate.

1. Circle the expressions that are polynomials.

| | |
|---|---|
| $-\sqrt{325}$ | $\frac{2}{5}p^3r - 3p^2q + \sqrt{2r}$ |
| $t^2 - s + 5$ | $\frac{5}{7}c^{15} + \frac{3}{14}c^{11} - 3\pi$ |
| $m^5n^4o^3p^2r$ | $x^2 + 3xy - \frac{2}{3x} + y^2$ |

2. Write m beside the monomial(s), b beside the binomial(s), and t beside the trinomial(s).

   a. ___ $w^5x^4$

   b. ___ $2x^2 - 36$

   c. ___ $\frac{1}{3}x^{17} + \frac{2}{3}x^{12} - \frac{1}{3}$

   d. ___ 27

   e. ___ $27x^3 - 2x^2y^3$

   f. ___ $x^2 + 3xy - \frac{2}{3}y^2$

3. Given the polynomial $3w^3 - 13w^2 + 7w^5 + 8w^8 - 2$, write the terms in descending order by degree.

4. Find:

   a. $(5x^3y - 8x^2y^2 + 3xy - y^3 + 13) + (-2xy + 6 + y^2 - 4y^3 - 2x^3y)$

   b. $(5x^3y - 8x^2y^2 + 3xy - y^3 + 13) - (-2xy + 6 + y^2 - 4y^3 - 2x^3y)$

5. Find: $x^3y^2w \cdot x^5yw^4$

6. Find: $n^2p^3(3n + 2n^3p^2 - 35p^4)$

7. Find: $21x^5y^2z^7 \div 14xyz$

8. Find: $(15t^3u^2v - 5t^5uv^2) \div 10tuv^2$

# ANSWERS

## Homework

**1.** $3\frac{1}{4}y^3 + 3y^2 - \sqrt{5}$ **3a.** $-4y^5 - 2y^3 + 3y + 2$

**b.** 5, 3, 1, 0 **c.** 5 **5.** $-4v^7 + v^3 + 6v^2 - 5v + 5$

**7.** $-7s^3t^3 + 7st^2 - s^2t + 2st - 13t + 9$

**9.** $2x^2y + 10xy^2 + 4y^3 + 3$

**11.** $4w^2yz + 3w^3 - 4wyz^2 + 6wyz - 4wy^2z + 3$

**13.** $x^3y^3z^3$ **15.** $-3t^4u^4v^{15}$ **17.** $10p^3r^4 + 5p^4r^5$

**19.** $\frac{3xw^5}{y}$ **21.** $\frac{3a^3d}{2b^5c^3}$ **23.** $\frac{4xy^2}{3} + \frac{5x^2y^3}{3}$ or $\frac{xy^2}{3}(4 + 5xy)$

## Practice Problems

**1.** $2xy + 5xz$; $9y^2 + 13yz - 8z^2$

**3a.** binomial **b.** binomial **c.** trinomial **d.** monomial **e.** trinomial

**5.** 8 **7.** 9 **9.** 7 **11.** 6 **13.** 84 **15.** $6x^2 + 11x - 8$

**17.** $15m^2n^3 + 2m^2n^2 - 7mn$ **19.** $15a^3b^2 + 4a^2b - ab^3$

**21.** $20xy^2z^3 - 30x^2yz^2 + 10x^3y^3z$ **23.** $4x^3 + 7x - 8$

**25.** $y^2 + 6xy + 4y$ **27.** $11a^5b^3 - 4a^4b - 9b$ **29.** $15y^5$

**31.** $-45a^9$ **33.** $28x^4y^8$ **35.** $-6w^2x^5y^3z^3$

**37.** $-6a^7b^7 + 10a^4b^5 - 12a^4b^2$

**39.** $20a^4b^2 + 10a^4b^3 - 35a^3b^4 - 15a^2b^3$

**41.** $12x^6y^3 - 28x^4y^5 + 8x^4y^4 - 4x^3y^4$

**43.** $5a^2b^5$ **45.** $\frac{8a^2b^3}{3c}$ **47.** $\frac{3x^3y^2z^5}{2w}$ **49.** $\frac{3n^5p^3}{2mq}$

**51.** $4a + 3a^3$ **53.** $\frac{7}{y} + 4x^3y$ **55.** $\frac{2}{x} - \frac{x^2z^3}{2y^2}$

## Practice Test

**1.** $t^2 - s + 5$, $m^5n^4o^3p^2r$, and $\frac{5}{7}c^{15} + \frac{3}{14}c^{11} - 3\pi$

**2.** $w^5x^4$ is a monomial

$2x^2 - 36$ is a binomial

$\frac{1}{3}x^{17} + \frac{2}{3}x^{12} - \frac{1}{3}$ is a trinomial

27 is a monomial

$27x^3 - 2x^2y^3$ is a binomial

$x^2 + 3xy - \frac{2}{3}y^2$ is a trinomial

**3.** $8w^8 + 7w^5 + 3w^3 - 13w^2 - 2$

**4a.** $3x^3y - 8x^2y^2 - 5y^3 + xy + y^2 + 19$

**b.** $7x^3y - 8x^2y^2 + 3y^3 + 5xy - y^2 + 7$

**5.** $x^8y^3w^5$

**6.** $3n^3p^3 + 2n^5p^5 - 35n^2p^7$

**7.** $\frac{3x^4yz^6}{2}$

**8.** $\frac{3t^2u}{2v} - \frac{1}{2}t^4$

# LESSON 6.3 – POLYNOMIAL OPERATIONS II

# OVERVIEW

***Here's what you'll learn in this lesson:***

***Multiplying Binomials***

*a. Multiplying binomials by the "FOIL" method*

*b. Perfect squares, product of the sum and difference of two terms*

***Multiplying and Dividing***

*a. Multiplying a polynomial by a polynomial*

*b. Dividing a polynomial by a polynomial*

Polynomials can be used to solve many types of problems. Some people might use a polynomial to create a household budget. A structural engineer might use a polynomial to find the wind force on a large building. Or, an automobile company might use a polynomial to find the average cost of manufacturing an airbag.

In this lesson, you will learn more about Polynomial Operations. You will multiply and divide polynomials which have more than one term.

# EXPLAIN

## MULTIPLYING BINOMIALS

### Summary

The multiplication of a binomial by a binomial can be simplified by using the "FOIL" method or by using patterns.

### Using the FOIL Method to Multiply Two Binomials

The FOIL method can be used to multiply any two binomials. The letters in the word "FOIL" show you the order in which to multiply.

The general format is:

$$(a+b)(c+d) = a \cdot c + a \cdot d + b \cdot c + b \cdot d$$

F + O + I + L

First + Outer + Inner + Last

To multiply two binomials using the FOIL method:

1. Multiply the First terms of the binomials.
2. Multiply the Outer terms (the terms next to the outer parentheses).
3. Multiply the Inner terms (the terms next to the inner parentheses).
4. Multiply the Last terms.
5. Add the terms. Be sure to combine like terms.

For example, to find: $(x-2)(x+5)$

1. Multiply the First terms: $x \cdot x$
2. Multiply the Outer terms: $x \cdot 5$
3. Multiply the Inner terms: $-2 \cdot x$
4. Multiply the Last terms: $-2 \cdot 5$
5. Add the terms. $x^2 + 5x - 2x - 10$

   $= x^2 + 3x - 10$

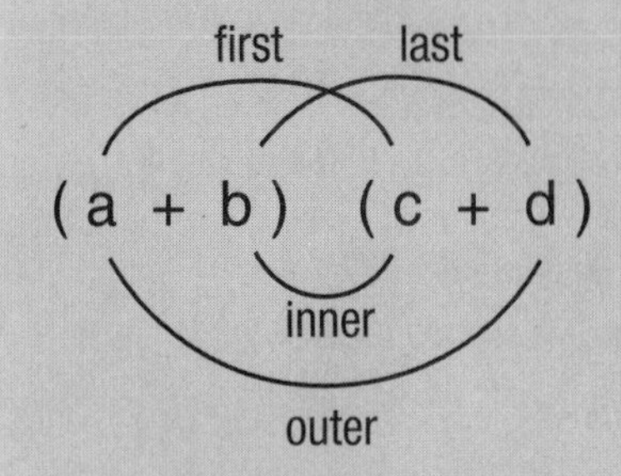

*This picture may help you remember how to use the FOIL method. Notice how the connecting lines form a face: F and L make the eyebrows, O makes the smile and I the nose.*

## Using Patterns to Multiply Two Binomials

Patterns can be used to find certain binomial products.

In general, when you square a binomial you can use one of these patterns:

$$(a + b)^2 = (a + b)(a + b) = a^2 + 2ba + b^2$$

$$(a - b)^2 = (a - b)(a - b) = a^2 - 2ba + b^2$$

These products are called perfect square trinomials.

Another binomial product that follows a pattern has the form:

$$(a + b)(a - b) = a^2 - b^2$$

This product is called a difference of two squares.

*If you forget the patterns, you can always use the FOIL method to figure them out:*

*F + O + I + L*

$(a + b)(a + b)$

$= a \cdot a + b \cdot a + b \cdot a + b \cdot b$

$= a^2 + 2ba + b^2$

$(a - b)(a - b)$

$= a \cdot a - b \cdot a - b \cdot a + (-b) \cdot (-b)$

$= a^2 - 2ba + b^2$

$(a + b)(a - b)$

$= a \cdot a - b \cdot a + b \cdot a - b \cdot b$

$= a^2 - b^2$

To use a pattern to find the product of two binomials:

1. Determine which pattern to use.
2. Determine which values to substitute for $a$ and $b$ in the pattern.
3. Substitute the values into the pattern.
4. Simplify.

For example, to find $(3x^2 + 4)(3x^2 + 4)$:

1. Determine which pattern to use.

   ✓ $(a + b)(a + b) = a^2 + 2ba + b^2$

   ___ $(a - b)(a - b) = a^2 - 2ba + b^2$

   ___ $(a + b)(a - b) = a^2 - b^2$

2. Determine the values to substitute for $a$ and $b$.

   $a = 3x^2,\ b = 4$

3. Substitute $3x^2$ for $a$ and 4 for $b$.

   $(a + b)^2 = a^2 + 2ba + b^2$

   $(3x^2 + 4)^2 = (3x^2)^2 + 2 \cdot 4 \cdot 3x^2 + 4^2$

4. Simplify.

   $= 9x^4 + 24x^2 + 16$

## Sample Problems

1. Use the FOIL method to find: $(x - 6y)(3x + 2y)$

- ☑ a. Multiply the First terms. $x \cdot 3x$
- ☐ b. Multiply the Outer terms. $x \cdot$ ____
- ☐ c. Multiply the Inner terms. ____ $\cdot 3x$
- ☐ d. Multiply the Last terms. ____ $\cdot$ ____
- ☐ e. Add the terms, combining like terms. = ________________

2. Use a pattern to find: $(t - 7)^2$

- ☑ a. Determine which pattern to use.
  - ___ $(a + b)^2 = a^2 + 2ba + b^2$
  - ✓ $(a - b)^2 = a^2 - 2ba + b^2$
  - ___ $(a + b)(a - b) = a^2 - b^2$
- ☐ b. Determine the values to substitute for $a$ and $b$. $a = t$, $b =$ ____
- ☐ c. Substitute these values. $(t - 7)^2 = t^2 - 2 \cdot$ ___ $\cdot\ t +$ ___
- ☐ d. Simplify. = ______________

3. Use a pattern to find: $(x + 5y)(x - 5y)$

- ☐ a. Determine which pattern to use.
  - ___ $(a + b)^2 = a^2 + 2ba + b^2$
  - ___ $(a - b)^2 = a^2 - 2ba + b^2$
  - ___ $(a + b)(a - b) = a^2 - b^2$
- ☑ b. Determine which values to substitute for $a$ and $b$. $a = x$, $b = 5y$
- ☐ c. Substitute these values. $(x + 5y)(x - 5y) =$ ____________
- ☐ d. Simplify. = ________

***Answers to Sample Problems***

1.
*b.* $2y$
*c.* $-6y$
*d.* $-6y$, $2y$
*e.* $3x^2 - 16xy - 12y^2$ *(in any order)*

2.
*b.* $7$
*c.* $7$; $7^2$ *or* $49$
*d.* $t^2 - 14t + 49$ *(in any order)*

3.
*a.* $(a + b)(a - b) = a^2 - b^2$
*c.* $x^2 - (5y)^2$
*d.* $x^2 - 25y^2$

# MULTIPYING AND DIVIDING

## Summary

### Multiplying Two Polynomials When Each Has More Than One Term

*In general, to multiply a polynomial by a polynomial when each has more than one term:*

*$(a + b)(c + d + e)$*

*$= a(c + d + e) + b(c + d + e)$*

*$= ac + ad + ae + bc + bd + be$*

You can multiply two polynomial each of which has more than one term.

To multiply a polynomial by a polynomial:

1. Distribute each term in the first polynomial to the second polynomial.
2. Distribute again to remove the parentheses.
3. Multiply each of the resulting terms.
4. Combine like terms.

For example, to find: $(x^2 + y)(3x^2 - 2y + xy)$

1. Distribute each term in the first polynomial to the second polynomial.
$= x^2(3x^2 - 2y + xy) + y(3x^2 - 2y + xy)$
2. Distribute again. $= (x^2)(3x^2) - x^2(2y) + (x^2)(xy) + y(3x^2) - y(2y) + y(xy)$
3. Multiply. $= 3x^4 - 2x^2y + x^3y + 3x^2y - 2y^2 + xy^2$
$= 3x^4 - 2x^2y + 3x^2y + x^3y - 2y^2 + xy^2$
4. Combine like terms. $= 3x^4 + x^2y + x^3y - 2y^2 + xy^2$

### Dividing a Polynomial with More Than One Term by Another Polynomial with More Than One Term

To divide a polynomial (dividend) by another polynomial (divisor) where each has more than one term, use long division.

Before you can divide, both polynomials should be arranged so their terms are in descending order by degree. To arrange the terms of a polynomial in descending order:

1. Determine the degree of each term by looking at the exponent of the variable(s).
2. Arrange the terms so they are in descending order by degree.

For example, to rearrange the terms of $x^3 - x + x^4$ in descending order:

1. Determine the degree of each term.
degree 3 degree 1 degree 4
$x^3 - x^1 + x^4$
2. Arrange the terms in descending order by degree. $x^4 + x^3 - x^1$

Once the terms of the polynomials are correctly arranged, you are ready to divide.

To divide a polynomial by a polynomial where each has more than one term:

1. Arrange the terms of each polynomial in descending order. In the dividend, write missing terms as $0x^r$ where $r$ is the exponent of the missing term.
2. Write the problem in long division form.
3. Divide the first term of the dividend by the divisor.
4. Multiply the divisor by the term you found in step (3).
5. Subtract the expression you found in step (4) from the dividend.
6. Continue dividing until the degree of the remainder is less than the degree of the divisor.
7. The answer is the expression that appears above the division sign plus the fraction $\frac{remainder}{divisor}$.
8. Check your division by multiplying the expression that appears above the division sign. Then add the remainder (not as a fraction).

For example, to find: $(x^2 + 3x^3 - 2) \div (x^2 + 2)$

1. Arrange the terms of the dividend in descending order. Include missing terms.

$$3x^3 + x^2 + 0x^1 - 2$$

2. Write the problem in long division form.

$$x^2 + 2\,\overline{)\,3x^3 + x^2 + 0x^1 - 2}$$

3. Divide the first term of the dividend by the divisor.
4. Multiply the divisor by the term in (3):
5. Subtract the expression you found in step (4) from the dividend.

$$\begin{array}{r} 3x \qquad\qquad\qquad \\ x^2 + 2\,\overline{)\,3x^3 + x^2 + 0x^1 - 2} \\ -(3x^3 \qquad + 6x) \quad \\ \hline x^2 - 6x - 2 \end{array}$$

6. Continue dividing until the degree of the remainder is less than the degree of the divisor.

$$\begin{array}{r} 3x + 1 \qquad\qquad \\ x^2 + 2\,\overline{)\,3x^3 + x^2 + 0x^1 - 2} \\ -(3x^3 \qquad + 6x) \quad \\ \hline x^2 - 6x - 2 \\ -(x^2 \qquad + 2) \\ \hline -6x - 4 \end{array}$$

7. Write your answer.

$$3x + 1 + \frac{-6x - 4}{x^2 + 2}$$

8. Check your answer by multiplying.

$$\begin{aligned} &(3x + 1)(x^2 + 2) + (-6x - 4) \\ &= 3x^3 + 6x + x^2 + 2 - 6x - 4 \\ &= 3x^3 + x^2 - 2 \end{aligned}$$

*Answers to Sample Problems*

c. $10tu^2, -2t^2u, -8u^3$ (in any order)

d. $3t^2u - t^3 + 6tu^2 - 8u^3$ (in any order)

c. 7

5

$21x - 21$

16

d. $5x^2, 5x, 7$

e. $5x, 7, 16$

Here's one way to do the check:

$(5x^2 + 5x + 7)(3x - 3) + 16$
$= 15x^3 - 15x^2 + 15x^2 - 15x + 21x - 21 + 16$
$= 15x^3 + 6x - 5$

## Sample Problems

1. Find: $(t + 2u)(5tu - t^2 - 4u^2)$

   ☑ a. Distribute each term in the first polynomial to each term in the second polynomial. $= t(5tu - t^2 - 4u^2) + 2u(5tu - t^2 - 4u^2)$

   ☑ b. Distribute again to remove parentheses. $= t \cdot 5tu - t \cdot t^2 - t \cdot 4u^2 + 2u \cdot 5tu - 2u \cdot t^2 - 2u \cdot 4u^2$

   ☐ c. Multiply each of the resulting terms. $= 5t^2u - t^3 - 4tu^2 +$ ____ $+$ ____ $+$ ____

   ☐ d. Combine like terms. $=$ ____________

2. Find: $(6x + 15x^3 - 5) \div (3x - 3)$

   ☑ a. Arrange the terms of the dividend in descending order. Include "missing" terms. $15x^3 + 0x^2 + 6x - 5$

   ☑ b. Write the division in long division form.

   ☐ c. Divide.

$$
\begin{array}{r}
5x^2 + 5x + \underline{\quad} \\
3x - 3\,\overline{)\,15x^3 + 0x^2 + 6x - 5} \\
-\underline{(15x^3 - 15x^2)} \\
15x^2 + 6x - 5 \\
-\underline{(15x^2 - 15x)} \\
21x - \underline{\quad} \\
-(\underline{\qquad}) \\
\underline{\quad}
\end{array}
$$

   ☐ d. Write the quotient. ____ $+$ ____ $+$ ____ $+ \frac{16}{3x - 3}$

   ☐ e. Check your division by multiplying the quotient by the divisor. $(5x^2 +$ ____ $+$ ____$)(3x - 3) +$ ____

# EXPLORE

## Sample Problems

On the computer, you found the products of two binomial factors. In particular, you found patterns such as a perfect square trinomial and the difference of two squares to help you multiply binomials without using the FOIL method. Below are some additional problems using these patterns.

1. Find $(x + y)^2 + (x - y)^2$.

   ☐ a. First find $(x + y)^2$. This follows the pattern $(a + b)^2 = a^2 + 2ab + b^2$. $(x + y)^2 =$ __________

   ☐ b. Then find $(x - y)^2$. This follows the pattern $(a - b)^2 = a^2 - 2ab + b^2$. $(x - y)^2 =$ __________

   ☐ c. Now combine terms.

   $(x + y)^2 + (x - y)^2 = (x^2 + 2xy + y^2) + (x^2 - 2xy + y^2)$

   $=$ __________

2. Find $(x + y)^2 - (x - y)^2$ first by using the perfect square trinomial patterns and then by using the pattern of a difference of two squares.

   ☐ a. First find $(x + y)^2$. This follows the pattern $(a + b)^2 = a^2 + 2ab + b^2$. $(x + y)^2 =$ __________

   ☐ b. Then find $(x - y)^2$. This follows the pattern $(a - b)^2 = a^2 - 2ab + b^2$. $(x - y)^2 =$ __________

   ☐ c. Now simplify.

   $(x + y)^2 - (x - y)^2 = (x^2 + 2xy + y^2) - (x^2 - 2xy + y^2)$

   $=$ __________

   ☐ d. Now solve the same problem using the pattern of a difference of two squares.

   Hint: Use the pattern $a^2 - b^2 = (a + b)(a - b)$. For this example, $a = (x + y)$ and $b = (x - y)$.

   $(x + y)^2 - (x - y)^2 = [(x + y) + (x - y)][(x + y) - (x - y)]$

   $= \quad 2x \quad \cdot$ _____

   $=$ _____

***Answers to Sample Problems***

1. *a.* $x^2 + 2xy + y^2$

   *b.* $x^2 - 2xy + y^2$

   *c.* $2x^2 + 2y^2$ *or* $2(x^2 + y^2)$

2. *a.* $x^2 + 2xy + y^2$

   *b.* $x^2 - 2xy + y^2$

   *c.* $4xy$

   *d.* $2y$

   $4xy$

**_Answers to Sample Problems_**

a. $10a^3b^2$, $10a^2b^3$, $5a^1b^4$

c. $3x$, $2y$, $3x$, $2y$, $3x$, $2y$

d. 216, 216, 96

Refer to this diagram of Pascal's triangle as you complete the additional exploration problems below.

$(a + b)^0$ — $1$

$(a + b)^1$ — $1a + 1b$

$(a + b)^2$ — $1a^2 + 2ab + 1b^2$

$(a + b)^3$ — $1a^3 + 3a^2b + 3ab^2 + 1b^3$

$(a + b)^4$ — $1a^4 + 4a^3b + 6a^2b^2 + 4ab^3 + 1b^4$

3. Expand $(a + b)^5$.

☐ a. Find the exponents and coefficients for each term.

$1a^5 + 5a^4b^1 + \_\_a^{\_}b^{\_} + \_\_a^{\_}b^{\_} + \_\_a^{\_}b^{\_} + 1b^5$

Hint: Use the coefficients for the expansion of $(a + b)^4$ to help you find the coefficient for each term in $(a + b)^5$.

$1a^4 + 4a^3b + 6a^2b^2 + 4ab^3 + 1b^4$

$1a^5 + 5a^4b + \_\_a^{\_}b^{\_} + \_\_a^{\_}b^{\_} + \_\_a^{\_}b^{\_} + 1b^5$

4. Use Pascal's triangle to find: $(3x + 2y)^4$

☑ a. Determine which row of Pascal's triangle will help.

$(3x + 2y)^4$

So use the row that expands $(a + b)^4$

☑ b. Write down the expansion for $(a + b)^4$.

$(a + b)^4 = 1a^4 + 4a^3b + 6a^2b^2 + 4ab^3 + 1b^4$

☐ c. Replace $a$ with $3x$ and $b$ with $2y$.

$1(3x)^4 + 4(\_\_)^3(\_\_) + 6(\_\_)^2(\_\_)^2 + 4(\_\_)(\_\_)^3 + 1(2y)^4$

☐ d. Simplify.

$= 81x^4 + \_\_x^3y + \_\_x^2y^2 + \_\_xy^3 + 16y^4$

# HOMEWORK

## Homework Problems

Circle the homework problems assigned to you by the computer, then complete them below.

### Explain
### Multiplying Binomials

1. Given $(2p + 3)(p - p^2)$, find the:

   First terms: ____ and ____

   Outer terms: ____ and ____

   Inner terms: ____ and ____

   Last terms: ____ and ____

2. Which pattern could you use to find each of the products (a) - (f) below? Write the appropriate pattern number next to each polynomial.

   a. ___ $(2x + 5y)^2$
   b. ___ $(2x + 5y)(2x - 5y)$
   c. ___ $(3t - 2)^2$
   d. ___ $(3t + 2)^2$
   e. ___ $(3t - 2)(3t - 2)$
   f. ___ $(2x^2 - 5y^3)^2$

   I. $(a + b)^2 = a^2 + 2ba + b^2$
   II. $(a - b)^2 = a^2 - 2ba + b^2$
   III. $(a + b)(a - b) = a^2 - b^2$

3. Given $(2s^3 + 5)^2$ and the pattern $(a + b)^2 = a^2 + 2ba + b^2$:

   a. What would you replace $a$ with in the pattern? _____

   b. What would you replace $b$ with? _____

4. Use a pattern to find: $(3s + 5)^2$

5. Use the FOIL method to find: $(4x - 2y)(3x + 6)$

6. Use a pattern to find: $(3t + 4u)(3t - 4u)$

7. Use patterns to find these products:

   a. $(3x^2 - 2)(3x^2 + 2)$

   b. $(3x^2 - 2)(3x^2 - 2)$

   c. $(3x^2 + 2)(3x^2 + 2)$

8. Find: $(5x^3 + 3y^2)^2$

9. A fish tank broke at the pet store where Angelina works, and part of the store was flooded. Since Angelina lost her measuring tape, she used a stick and her handspan to figure out the approximate size of the flooded area. If $s$ equals the length of the stick and $h$ equals the width of her handspan, these are the measurements:

   length of flooded space $= 13s + 2h$

   width of flooded space $= 13s - 2h$

   area of flooded space $= (13s + 2h)(13s - 2h)$

   Simplify the equation for the area by multiplying the binomials.

10. The owner of the pet store where Angelina works wants to replace the tile covering the entire floor, not just the flooded area. If the length of the entire floor is $250s - 3h$ and the width is $98s + h$, what is the area of the floor in terms of $s$ and $h$?
    Hint: area = length · width.

11. Find: $(13x^2y^2 - 10x^3)(7x^2y^2 - 6x^3)$

12. Find:

    a. $\left(\frac{1}{2}x^3 - \frac{2}{3}y^5\right)\left(\frac{1}{2}x^3 - \frac{2}{3}y^5\right)$

    b. $\left(\frac{1}{2}x^3 - \frac{2}{3}y^5\right)\left(\frac{1}{2}x^3 + \frac{2}{3}y^5\right)$

    c. $\left(\frac{1}{2}x^3 + \frac{2}{3}y^5\right)\left(\frac{1}{2}x^3 + \frac{2}{3}y^5\right)$

### Multiplying and Dividing

13. Find: $(x + 2)(3x + 4xy + 1)$

14. Find: $(p^2 + 2r + 2)(3r^4 - 2p^4)$

15. Find: $(x + y + 1)(x - y)$

16. Find: $(2t + u)(t + 2u - 1)$

17. Angelina is cleaning the windows of the guinea pig case at the pet store where she works. The surface area of the outside of the windows can be described as follows:

    surface area = $2(x + 3)(x - 2) + 2(x - 2)(x - 3)$

    Simplify this equation by multiplying the polynomials.

18. The pet store where Angelina works sells an exercise arena for guinea pigs, consisting of two spheres connected by a tube. The volume of the exercise arena can be described by this equation:

    volume = $4\pi r^3 + 3\pi(r^2 + 2r + 4)(r + 2) + \pi r(r - 5)(r - 5)$

    Simplify the equation by multiplying the polynomials.

19. Find: $(12x^3 - 2x^2 - 7x)\left(4x^2 - \frac{10}{3}x - \frac{1}{3} + \frac{7}{12x^3 - 2x^2 - 7x}\right)$

20. Find: $\left(\frac{1}{3}t^2 + \frac{2}{3}v^3\right)\left(\frac{1}{3}t^2 + \frac{2}{3}v^3\right)\left(\frac{1}{3}t^2 + \frac{2}{3}v^3\right)$

21. Find: $(3x^2 + 2x - 1) \div (x + 3)$

22. Here is how Tony answered a question on his algebra test. $(12x^3 - 17x^2 + 3) \div (3x - 2) = 4x^2 - 3x - 2$ remainder $-1$.

    Is his answer right or wrong? Why? Circle the most appropriate response.

    His answer is right.

    His answer is wrong. When doing the long division, he sometimes added negative terms rather than subtracting them. The right answer is $4x^2 + 2x - 1$.

    His answer is wrong. He did not include missing terms in the quotient. The right answer is $0x^3 + 4x^2 - 3x - 2$ remainder $-1$.

    His answer is wrong. He did not put the remainder over the dividend. The right answer is $4x^2 - 3x - 2 + \frac{-1}{3x - 2}$.

23. Find: $(15x^5 + x^4 + 5x^2 - 2) \div (3x^2 + 2x)$

24. Find: $(8y^6 + 4y^4 - 10y^3 - 12) \div (2y^3 - 2y + 4)$

## Explore

25. Find: $(3a - 1)(3a + 1)$

26. Use the table below to find a general form for multiplying two polynomials: $(ax^2 + bx + c)(dx - e)$

| terms | $dx$ | $-e$ |
|---|---|---|
| $ax^2$ | | |
| $bx$ | | |
| $c$ | | |

$(ax^2 + bx + c)(dx - e) =$

27. Use the table below to find the general form for a difference of two squares: $(a + b)(a - b)$. Then use this pattern to find $(2x + 3y)(2x - 3y)$.

| terms | $a$ | $-b$ |
|---|---|---|
| $a$ | | |
| $b$ | | |

$(a + b)(a - b) =$

$(2x + 3y)(2x - 3y) =$

28. Find: $(x^2 + 3y)^2$

29. Use the table below to find the general form for a perfect square trinomial: $(a - b)(a - b)$. Then use the pattern to find $(2t^3 - 4u^2)(2t^3 - 4u^2)$.

| terms | $a$ | $-b$ |
|---|---|---|
| $a$ | | |
| $-b$ | | |

$(a - b)(a - b) =$

$(2t^3 - 4u^2)(2t^3 - 4u^2) =$

30. Use the table below to find the general form for a perfect square trinomial: $(a + b)(a + b)$. Then use this general form to find $(x^2 + 3y)(x^2 + 3y)$.

| terms | $a$ | $b$ |
|---|---|---|
| $a$ | | |
| $b$ | | |

$(a + b)(a + b) =$

$(x^2 + 3y)(x^2 + 3y) =$

# APPLY

## Practice Problems

Here are some additional practice problems for you to try.

### Multiplying Binomials

1. Find: $(a + 2)(a + 5)$
2. Find: $(m - 3)(m - 7)$
3. Find: $(x - 4)(x - 11)$
4. Find: $(3b + 2)(b - 6)$
5. Find: $(5y - 8)(y + 3)$
6. Find: $(6t + 1)(t - 7)$
7. Find: $(4a + 3b)(2a + 5b)$
8. Find: $(3m - 4n)(7m + 2n)$
9. Find: $(6y + 5x)(3y - x)$
10. Find: $(p + 9)(p + 9)$
11. Find: $(x + 3)(x + 3)$
12. Find: $(3z + 2)(3z + 2)$
13. Find: $(5q + 3)(5q + 3)$
14. Find: $(4x + 1)(4x + 1)$
15. Find: $(z - 5)(z - 5)$
16. Find: $(m - 11)(m - 11)$
17. Find: $(t - 6)(t - 6)$
18. Find: $(3x - 2y)(3x - 2y)$
19. Find: $(4a - 7c)(4a - 7c)$
20. Find: $(5r - 8s)(5r - 8s)$
21. Find: $(5m + n)(5m - n)$
22. Find: $(a + 7b)(a - 7b)$
23. Find: $(2x + y)(2x - y)$
24. Find: $(3y + 8)(3y - 8)$
25. Find: $(5x + 3)(5x - 3)$
26. Find: $(m + 12n)(m - 12n)$
27. Find: $(2a + 7b)(2a - 7b)$
28. Find: $(x + 7y)(x - 7y)$

### Multiplying and Dividing

29. Find: $(4a - 3b)(2a - 7b)$
30. Find: $(3x + 5)(y + 8)$
31. Find: $(6m - 5n)(3m + 4n)$
32. Find: $(8y + 3z)(2y - 9z)$
33. Find: $(7x - 4)(2y + 3)$
34. Find: $(a + 2b)(a^2 + 6a - 3b)$
35. Find: $(3mn - n)(m^2 - 3n + 4m)$
36. Find: $(2xy - y)(x^2 + 5y - 6x)$
37. Find: $(3ab + 4b)(7a^2 + 3b - 4a)$
38. Find: $(7uv - 3v)(2u^2 - 5v + 8u)$
39. Find: $(5xy + 2y)(2x^2 - 6y + 3x)$
40. Find: $(3a^2 - 4b^2)(2a^3 + 5a^2b - 11ab - b)$
41. Find: $(5m^2n + 3n)(4m^3 - 3m^2n + 8mn^2 - 3n^2)$
42. Find: $(7x^2y + 2y)(3x^3 - 6x^2y + 8xy + y)$
43. Find: $(x^3 + x^2 - 13x + 14) \div (x - 2)$
44. Find: $(x^3 + 11x^2 + 22x - 24) \div (x + 4)$
45. Find: $(x^3 + 10x^2 + 23x + 6) \div (x + 3)$
46. Find: $(x^3 + 7x^2 - 36) \div (x + 6)$
47. Find: $(x^3 - 26x + 5) \div (x - 5)$
48. Find: $(3x^3 + 17x^2 - 58x + 40) \div (3x - 4)$
49. Find: $(4x^3 + 4x^2 - 13x + 5) \div (2x + 5)$
50. Find: $(2x^3 + 7x^2 - x - 2) \div (2x + 1)$
51. Find: $(4x^3 + 7x^2 - 14x + 6) \div (4x - 1)$
52. Find: $(2x^3 - 9x^2 + 12x - 8) \div (2x + 3)$
53. Find: $(3x^3 + 14x^2 + 11x - 8) \div (3x + 2)$
54. Find: $(6x^3 - 7x^2 - 34x + 35) \div (2x - 5)$
55. Find: $(10x^3 - 26x^2 - 7x + 2) \div (5x + 2)$
56. Find: $(8x^3 - 18x^2 + 25x - 12) \div (4x - 3)$

# EVALUATE

## Practice Test

Take this practice test to be sure that you are prepared for the final quiz in Evaluate.

1. Use the FOIL method to find: $(2x^2 + 3xy)(3x^3y - 2)$

2. Use a pattern to find: $(2x - 3y)^2$

3. Find: $(2x + 3y)^2$

4. Use a pattern to find: $(2x - 3y)(2x + 3y)$

5. Find: $(3x - 2)(5x^2 + 8x - 2)$

6. Find: $(3p^2 + 4r^4 - 5)(3r^4 - 6p^2 + 2)$

7. Find: $(6t^3 + 5t^2 - 3t + 1) \div (2t + 1)$

8. Find: $(8x^3 + 6x - 2) \div (4x + 2)$

9a. Find: $(a^3 - a^5)(a + a^2)$

b. What is the degree of the resulting polynomial?

10. Find: $(5y^4 - 2y^2 + y)(3y^2 - y + 2)$

11. Use the table in Figure 6.3.1 to find:
$(2x^3 - 3x + 7)(5x^4 + 8)$

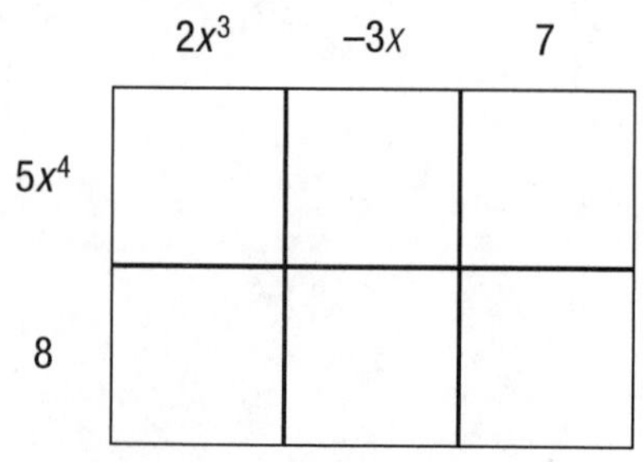


*Figure 6.3.1*

12. Use the table in Figure 6.3.2 to find:
$(5x^4 - 7x^3 + 7x^2 - 8x)(x^2 + 1)$

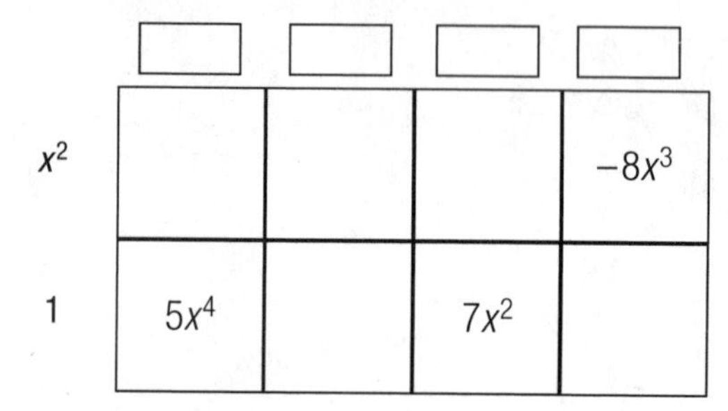


*Figure 6.3.2*

# ANSWERS

## Homework

**1.** First terms: $2p$ and $p$
Outer terms: $2p$ and $-p^2$
Inner terms: 3 and $p$
Last terms: 3 and $-p^2$

**3.** $2s^3, 5$ **5.** $12x^2 + 24x - 6yx - 12y$

**7a.** $9x^4 - 4$ **b.** $9x^4 - 12x^2 + 4$ **c.** $9x^4 + 12x^2 + 4$

**9.** $169s^2 - 4h^2$ **11.** $91x^4y^4 - 148x^5y^2 + 60x^6$

**13.** $3x^2 + 4x^2y + 7x + 8xy + 2$

**15.** $x^2 + x - y^2 - y$ or $x(x + 1) - y(y + 1)$ **17.** $4x^2 - 8x$

**19.** $48x^5 - 48x^4 - \frac{76}{3}x^3 + 24x^2 + \frac{7}{3}x + 7$

**21.** $3x - 7 + \frac{20}{x+3}$ **23.** $5x^3 - 3x^2 + 2x + \frac{1}{3} - \frac{\frac{2}{3x} + 2}{3x^2 + 2x}$

**25.** $9a^2 - 1$ **27.** $a^2 - b^2, 4x^2 - 9y^2$

**29.** $a^2 - 2ab + b^2, 4t^6 - 16u^2t^3 + 16u^4$

## Practice Problems

**1.** $a^2 + 7a + 10$ **3.** $x^2 - 15x + 44$ **5.** $5y^2 + 7y - 24$

**7.** $8a^2 + 26ab + 15b^2$ **9.** $18y^2 + 9xy - 5x^2$

**11.** $x^2 + 6x + 9$ **13.** $25q^2 + 30q + 9$ **15.** $z^2 - 10z + 25$

**17.** $t^2 - 12t + 36$ **19.** $16a^2 - 56ac + 49c^2$ **21.** $25m^2 - n^2$

**23.** $4x^2 - y^2$ **25.** $25x^2 - 9$ **27.** $4a^2 - 49b^2$

**29.** $8a^2 - 34ab + 21b^2$ **31.** $18m^2 + 9mn - 20n^2$

**33.** $14xy + 21x - 8y - 12$

**35.** $3m^3n + 11m^2n - 4mn - 9mn^2 + 3n^2$

**37.** $21a^3b + 16a^2b + 9ab^2 - 16ab + 12b^2$

**39.** $10x^3y - 30xy^2 + 19x^2y - 12y^2 + 6xy$

**41.** $20m^5n - 15m^4n^2 + 40m^3n^3 - 15m^2n^3 + 12m^3n - 9m^2n^2 + 24mn^3 - 9n^3$

**43.** $x^2 + 3x - 7$ **45.** $x^2 + 7x + 2$

**47.** $x^2 + 5x - 1$ **49.** $2x^2 - 3x + 1$

**51.** $x^2 + 2x - 3$ remainder 3 or $x^2 + 2x - 3 + \frac{3}{4x-1}$

**53.** $x^2 + 4x + 1$ remainder $-10$ or $x^2 + 4x + 1 - \frac{10}{3x+2}$

**55.** $2x^2 - 6x + 1$

## Practice Test

**1.** $(2x^2)(3x^3y) + (2x^2)(-2) + (3xy)(3x^3y) + (3xy)(-2)$

**2.** $4x^2 - 12xy + 9y^2$ **3.** $4x^2 + 12xy + 9y^2$

**4.** $4x^2 - 9y^2$ **5.** $15x^3 + 14x^2 - 22x + 4$

**6.** $12r^8 - 15p^2r^4 - 18p^4 - 7r^4 + 36p^2 - 10$

**7.** $3t^2 + t - 2 + \frac{3}{2t+1}$ **8.** $2x^2 - x + 2 + \frac{-6}{4x+2}$

**9a.** $a^4 + a^5 - a^6 - a^7$
**b.** The degree of the resulting polynomial is 7.

**10.** $15y^6 - 5y^5 + 4y^4 + 5y^3 - 5y^2 + 2y$

**11.**

| | $2x^3$ | $-3x$ | $7$ |
|---|---|---|---|
| $5x^4$ | $10x^7$ | $-15x^5$ | $35x^4$ |
| $8$ | $16x^3$ | $-24x$ | $56$ |

$10x^7 - 15x^5 + 35x^4 + 16x^3 - 24x + 56.$

**12.**

| | $5x^4$ | $-7x^3$ | $7x^2$ | $-8x$ |
|---|---|---|---|---|
| $x^2$ | $5x^6$ | $-7x^5$ | $7x^4$ | $-8x^3$ |
| $1$ | $5x^4$ | $-7x^3$ | $7x^2$ | $-8x$ |

# TOPIC 6 CUMULATIVE ACTIVITIES

## CUMULATIVE REVIEW PROBLEMS

These problems combine all of the material you have covered so far in this course. You may want to test your understanding of this material before you move on to the next topic. Or you may wish to do these problems to review for a test.

1. Find:

   a. $2^7 \cdot 2^9$  b. $\frac{x^{12}}{x^5}$  c. $(a^5b^2)^4$

2. Solve $-3 \leq 6 + 2y < 4$ for $y$, then graph its solution on the number line below.

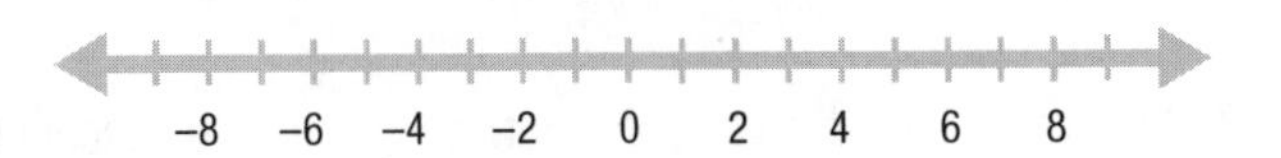


3. Find the equation of the line through the point (3, 7) with slope $-\frac{2}{7}$:

   a. in point-slope form.

   b. in slope-intercept form.

   c. in standard form.

4. Solve this system:

$$x + 2y = 5$$
$$x - 2y = -13$$

5. The difference of two numbers is –32. The sum of three times the smaller number and twice the larger number is 134. What are the two numbers?

6. Circle the true statements.

   The GCF of two numbers that have no factors in common is 1.

   $\frac{2}{9} - \frac{1}{5} = \frac{1}{4}$

   The LCM of 4 and 8 is 4.

   $3^2(4 + 2) = 9(4 + 2)$

   $\frac{1}{2} + \frac{1}{3} = \frac{5}{6}$

7. Write the equation of the line through the point (20, –9) with slope $-\frac{8}{5}$:

   a. in point-slope form.

   b. in slope-intercept form.

   c. in standard form.

8. Graph the system of inequalities below to find its solution.

$$2x + y \geq 3$$
$$x - y < 4$$

9. Find:

   a. $3^0$  b. $-3^0$  c. $(-3)^0$

10. Graph the inequality $4x + y \leq 6$.

11. Find: $15x^3y^8z^5 \div 10xy^4z^{11}$

12. Lisa emptied a vending machine and got a total of 279 quarters and dimes worth $57.30. How many quarters did she get?

13. Find the slope and $y$-intercept of the line $4x - y = 7$.

14. Evaluate the expression $3x^2 - 4xy + 2y$ when $x = 3$ and $y = -5$.

15. Solve $-6 < 4 + 2x < -2$ for $x$.

16. Find:

   a. $-5x^0 + y^2$

   b. $\left(\frac{a^3 \cdot b^7 \cdot c}{b^4 \cdot c^2}\right)^3$

   c. $b^4 \cdot b^2 \cdot b \cdot b^6$

17. Circle the true statements.

   $4(3 - 5) = 4 \cdot 3 - 5$

   $\frac{26}{117} = \frac{2}{9}$

   The LCM of 72 and 108 is 36.

   The GCF of 72 and 108 is 36.

   $\frac{4}{7} - \frac{2}{3} = -\frac{2}{4}$

18. Write the equation of the line through the point (5, 2) with slope $-\frac{7}{3}$:

    a. in point-slope form.

    b. in slope-intercept form.

    c. in standard form.

19. Graph the line $y = 6$.

20. Find: $(11p^2 - 3pr - 6r)(3p - 9r)$

21. Solve this system:

$$4x - y = 9$$
$$6x + 5y = -6$$

22. Find the equation of the line that is **parallel** to the line $x + 3y = 4$ and passes through the point (2, 2).

23. Find the equation of the line that is **perpendicular** to the line $x + 3y = 4$ and passes through the point (2, 2).

24. Graph the system of inequalities below to find its solution.

$$y < 2x + 3$$
$$4x - y \geq 1$$

25. Find the slope of the line perpendicular to the line through the points (8, 9) and (6, –4).

26. Find: $(x^3 + 5x^2 + x - 10) \div (x + 2)$

27. Solve $2y + 5 = 4(\frac{1}{2}y + 2)$ for $y$.

28. Find: $3xy(x^2y - 4)$

Use Figure 6.1 to answer questions 29, 30, and 31.

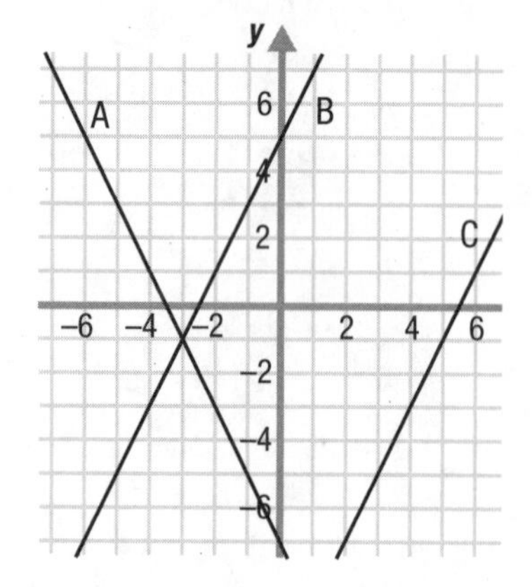


*Figure 6.1*

29. Which two lines form a system that has a solution of (–3, –1)?

30. Which two lines form a system that has no solution?

31. Which two lines form a system that has a solution that is not shown on the grid?

32. Find:

    a. $(x^2yz^3)^4$

    b. $\frac{x^5y^3}{xy^6}$

    c. $(x^5)^9$

33. Evaluate the expression $-7a^4 + 3ab^2 + b - 4$ when $a = -2$ and $b = 5$.

34. Find the slope of the line through the points (9, –4) and (2, 7).

35. Graph the system of inequalities below to find its solution.

$$9x - 4y < 20$$
$$9x - 4y \leq 8$$

36. Graph the inequality $\frac{5}{2}x - y \geq 2$.

37. Last year Manuel split \$2565 between his savings account, which paid 5% in interest, and his checking account, which paid 3.5% in interest. If he earned a total of \$113.49 in interest, how did he split his money between the two accounts?

38. Graph the inequality $\frac{2}{3}x - \frac{1}{3}y \geq 2$.

39. Next to each polynomial below, write whether it is a monomial, a binomial, or a trinomial.

    a. $2 + x$

    b. $8ab^3 - 9abc + 1$

    c. $3x^7yz^5$

    d. $a^5b^2c^3d - 103a^7cd^4$

    e. $10$

    f. $a + 7b - 4c$

40. Solve $-8(1 - \frac{1}{2}x) = 4(x - 2)$ for $x$.

41. Find: $(4a^2b + 3a - 9b) + (7a + 2b - 8a^2b)$

42. The perimeter of a square is the same as the perimeter of a regular hexagon. If each side of the square is 7 feet longer than each side of the hexagon, what is the perimeter of each figure?

# ANSWERS

## Cumulative Review Problems

**1a.** 216 **b.** $x^7$ **c.** $a^{20}b^8$ **3a.** $y - 7 = -\frac{2}{7}(x - 3)$

**b.** $y + -\frac{2}{7}x + \frac{55}{7}$ **c.** $2x + 7y = 55$

**5.** The numbers are 14 and 46. **7a.** $y + 9 = -\frac{8}{5}(x - 20)$

**b.** $y = -\frac{8}{5}x + 23$ **c.** $8x + 5y = 115$ **9a.** 1 **b.** –1 **c.** 1

**11.** $\frac{3x^2y^4}{2z^6}$ **13.** slope = 4, $y$–intercept = (0, –7)

**15.** $-5 < x < -3$

**17.** $\frac{26}{117}$, The LCM of 72 and 108 is 36, The GCF of 72 and 108 is 36

**19.**

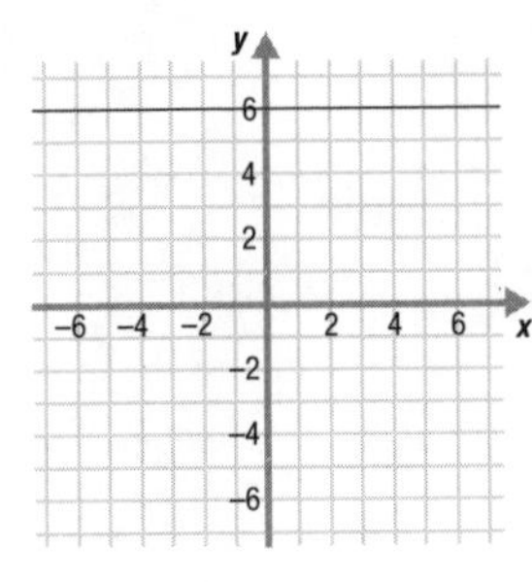

**21.** $\left(\frac{3}{2}, -3\right)$ **23.** $y = 3x - 4$ **25.** $-\frac{2}{13}$

**27.** No solution for $y$ **29.** A and B **31.** A and C **33.** –241

**35.**

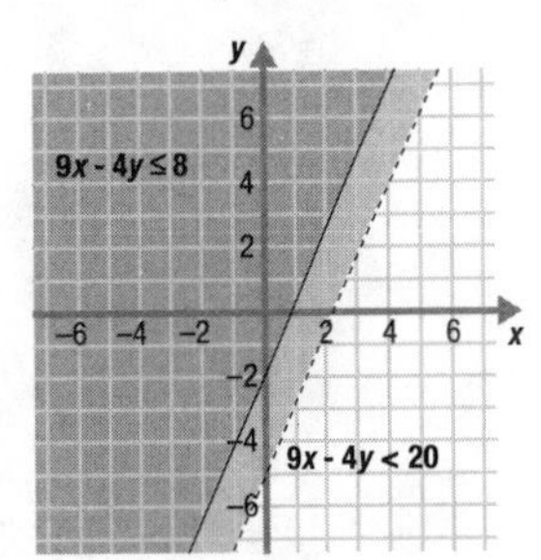


**37.** Manuel split his money by putting $984 in his checking account and $1581 in his saving account.

**39a.** binomial **b.** trinomial **c.** monomial **d.** binomial **e.** monomial **f.** trinomial

**41.** $-4a^2b + 10a - 7b$

# TOPIC 6 INDEX

# LESSON 7.1 – FACTORING POLYNOMIALS I



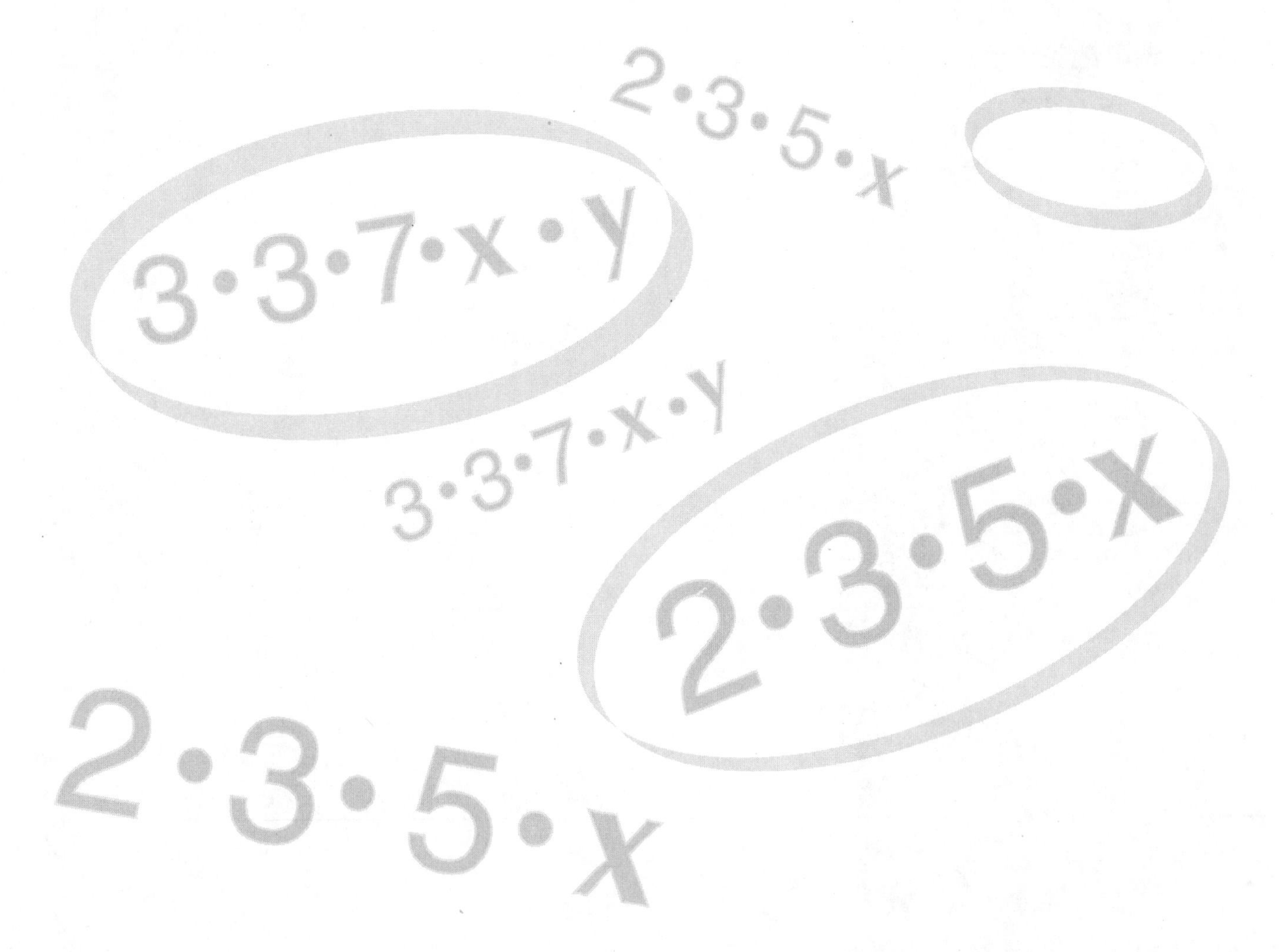

# OVERVIEW

***Here's what you'll learn in this lesson:***

***Greatest Common Factor***

*a. Finding the greatest common factor (GCF) of a set of monomials*

*b. Factoring a polynomial by finding the GCF when the GCF is a monomial*

***Grouping***

*a. Factoring a polynomial by finding the GCF when the GCF is a binomial*

*b. Factoring a polynomial with four terms by grouping*

You have learned how to multiply polynomials. Now you will learn how to factor them. When you factor a polynomial, you write it as the product of other polynomials.

In this lesson you will learn several different techniques for factoring polynomials.

# EXPLAIN

## GREATEST COMMON FACTOR

### Summary

#### Factoring Polynomials

You already know how to factor numbers by writing them as the product of other numbers. Now you will learn how to factor polynomials by writing them as the product of other polynomials.

#### Finding the GCF of a Collection of Monomials

To find the GCF of a collection of monomials:

1. Factor each monomial into its prime factors.
2. List each common prime factor the **smallest** number of times it appears in any factorization.
3. Multiply all the prime factors in the list.

For example, to find the GCF of the monomials $16x^2y^2$, $4x^3y^2$, and $12xy^4$:

1. Factor each monomial into its prime factors.

$$16x^2y^2 = 2 \cdot 2 \cdot 2 \cdot 2 \cdot x \cdot x \cdot y \cdot y$$

$$4x^3y^2 = 2 \cdot 2 \cdot x \cdot x \cdot x \cdot y \cdot y$$

$$12xy^4 = 2 \cdot 2 \cdot 3 \cdot x \cdot y \cdot y \cdot y \cdot y$$

2. List each common prime factor the **smallest** number of times it appears in any factorization. $2, 2, x, y, y$
3. Multiply all the prime factors in the list. $\text{GCF} = 2 \cdot 2 \cdot x \cdot y \cdot y = 4xy^2$

#### Factoring a Polynomial By Finding The Greatest Common Factor

One way to factor a polynomial is to find the greatest common factor of its monomial terms. Here are the steps:

1. Identify the monomial terms of the polynomial.
2. Factor each monomial term.
3. Find the GCF of the monomial terms.
4. Rewrite each term of the polynomial using the GCF.
5. Factor out the GCF.
6. Use the distributive property to check your factoring.

*Remember that a monomial is a polynomial with only one term. For example: $14x^5y^3$, 32, 6x, and 9xyz are monomials; but $12x^5y + 1$ and $14y + 3x$ are not monomials.*

*Before deciding if a polynomial is a monomial, binomial, etc., be sure you first combine any like terms and apply the distributive property, if possible.*

*The GCF of a collection of monomials is the GCF of the coefficients of all the monomials multiplied by the smallest power of each variable in all the monomials.*

*The GCF of a collection of monomials evenly divides each monomial in the collection.*

$$\frac{16x^2y^2}{4xy^2} = 4x$$

$$\frac{4x^3y^2}{4xy^2} = x^2$$

$$\frac{12xy^4}{4xy^2} = 3y^2$$

For example, to factor the polynomial $6x^2y^2 + 8y^2$:

1. Identify the terms of the polynomial. $6x^2y^2, 8y^2$
2. Factor each monomial term. $6x^2y^2 = 2 \cdot 3 \cdot x \cdot x \cdot y \cdot y$
   $8y^2 = 2 \cdot 2 \cdot 2 \cdot y \cdot y$
3. Find the GCF of the monomial terms. $\text{GCF} = 2 \cdot y \cdot y = 2y^2$
4. Rewrite each term of the polynomial using the GCF $6x^2y^2 = 2y^2 \cdot 3x^2$
   $8y^2 = 2y^2 \cdot 4$
5. Factor out the GCF. $6x^2y^2 + 8y^2 = (2y^2)(3x^2 + 4)$
6. Use the distributive property to check your factoring. Is $6x^2y^2 + 8y^2 = (2y^2)(3x^2 + 4)$ ?
   Is $6x^2y^2 + 8y^2 = (2y^2)(3x^2) + (2y^2)(4)$?
   Is $6x^2y^2 + 8y^2 = 6x^2y^2 + 8y^2$ ? Yes.

## Sample Problems

1. Find the GCF of $9x^3y$, $3xy^4$, and $6y^2$.

   ☐ a. Factor each monomial into its prime factors.
   $9x^3y = 3 \cdot 3 \cdot x \cdot x \cdot x \cdot y$
   $3xy^4 =$ ____________
   $6y^2 =$ ____________

   ☐ b. List each common factor the smallest number of times it appears in any factorization. ____________

   ☐ c. Multiply all the prime factors in the list. GCF = ________

2. Factor: $6x^4y^4 + 30x^2y^3 + 10x^5y^2$

   ☑ a. Find the terms of the polynomial. $6x^4y^4$, $30x^2y^3$, and $10x^5y^2$

   ☐ b. Factor each monomial.
   $6x^4y^4 =$ ____________
   $30x^2y^3 =$ ____________
   $10x^5y^2 =$ ____________

   ☐ c. Find the GCF of the monomial terms. GCF = ________

   ☐ d. Rewrite each term of the polynomial using the GCF.
   $6x^4y^4 =$ ____________
   $30x^2y^3 =$ ____________
   $10x^5y^2 =$ ____________

   ☐ e. Factor out the GCF. = (________)(________)

   ☐ f. Use the distributive property to check your factoring.

***Answers to Sample Problems***

*a.* $3 \cdot x \cdot y \cdot y \cdot y \cdot y$
$2 \cdot 3 \cdot y \cdot y$

*b.* $3, y$

*c.* $3y$

*b.* $2 \cdot 3 \cdot x \cdot x \cdot x \cdot x \cdot y \cdot y \cdot y \cdot y$
$2 \cdot 3 \cdot 5 \cdot x \cdot x \cdot y \cdot y \cdot y$
$2 \cdot 5 \cdot x \cdot x \cdot x \cdot x \cdot x \cdot y \cdot y$

*c.* $2x^2y^2$

*d.* $(2x^2y^2)(3x^2y^2)$
$(2x^2y^2)(15y)$
$(2x^2y^2)(5x^3)$

*e.* $(2x^2y^2)(3x^2y^2 + 15y + 5x^3)$

*f.* $(2x^2y^2)(3x^2y^2 + 15y + 5x^3)$
$= (2x^2y^2)(3x^2y^2) + (2x^2y^2)(15y) + (2x^2y^2)(5x^3)$
$= 6x^4y^4 + 30x^2y^3 + 10x^5y^2$

# GROUPING

## Summary

### Factoring a Polynomial by Finding the Binomial GCF

You have already learned how to factor a polynomial when the GCF of the terms of the polynomial is a monomial. You can use the same steps to factor a polynomial when the GCF of the terms is a binomial.

There are six steps in this procedure:

1. Identify the terms of the polynomial.
2. Factor each term.
3. Find the GCF of the terms.
4. Rewrite each term of the polynomial using the GCF.
5. Factor out the GCF.
6. Check your answer.

For example, to factor the polynomial $5(3x+2) + x^2(3x+2)$:

1. Identify the terms of the polynomial. $5(3x+2)$ and $x^2(3x+2)$
2. Factor each term. (Here each term is already factored.)
   $5(3x+2) = 5 \cdot (3x+2)$
   $x^2(3x+2) = x^2 \cdot (3x+2)$
3. Find the GCF of the terms. $\text{GCF} = 3x+2$
4. Rewrite each term of the polynomial using the GCF.
   $5(3x+2) = 5 \cdot (3x+2)$
   $x^2(3x+2) = x^2 \cdot (3x+2)$
5. Factor out the GCF.
   $$5(3x+2) + x^2(3x+2) = 5 \cdot (3x+2) + x^2 \cdot (3x+2)$$
   $$= (3x+2)(5+x^2)$$
6. Check your answer.

   Is $(3x+2)(5+x^2) = 5(3x+2) + x^2(3x+2)$?

   Is $(3x+2)(5) + (3x+2)(x^2) = 5(3x+2) + x^2(3x+2)$ ? Yes.

### Factoring By Grouping

Sometimes the GCF of the terms of a polynomial is 1.

For example find the GCF of the terms of $3x^2 + 9 + bx^2 + 3b$:

$$3x^2 = 1 \cdot 3 \cdot x \cdot x$$

$$9 = 1 \cdot 3 \cdot 3$$

$$bx^2 = 1 \cdot b \cdot x \cdot x$$

$$3b = 1 \cdot 3 \cdot b$$

You see that the GCF of the 4 terms $3x^2$, 9, $bx^2$, and $3b$ is 1. If you try to use the GCF to factor $3x^2 + 9 + bx^2 + 3b$ you get the following factorization:

$$3x^2 + 9 + bx^2 + 3b = 1 \cdot (3x^2 + 9 + bx^2 + 3b)$$

This isn't very interesting!

To factor the polynomial $3x^2 + 9 + bx^2 + 3b$ you need a technique other than finding the GCF of the terms. One such technique is called factoring by grouping. This procedure has 5 steps:

1. Factor each term.
2. Group terms with common factors.
3. Factor out the GCF in each grouping.
4. Factor out the binomial GCF of the polynomial.
5. Check your answer.

For example, use this technique to factor the polynomial $3x^2 + 9 + bx^2 + 3b$:

1. Factor each term.

$$3x^2 = 3 \cdot x \cdot x$$
$$9 = 3 \cdot 3$$
$$bx^2 = b \cdot x \cdot x$$
$$3b = 3 \cdot b$$

*This isn't the only way to group the terms. For example, you could also have grouped the terms like this:*

*$(3x^2 + bx^2) + (9 + 3b)$*

*Try it; you'll get the same answer.*

2. Group terms with common factors.

$$= 3x^2 + 9 + bx^2 + 3b$$
$$= (\mathbf{3} \cdot x \cdot x + \mathbf{3} \cdot 3) + (\boldsymbol{b} \cdot x \cdot x + 3 \cdot \boldsymbol{b})$$

3. Factor out the GCF in each grouping.

$$= \mathbf{3}(x \cdot x + 3) + \boldsymbol{b}(x \cdot x + 3)$$

4. Factor out the binomial GCF of the polynomial.

$$= \mathbf{3}(x^2 + 3) + \boldsymbol{b}(x^2 + 3)$$
$$= (\mathbf{3} + \boldsymbol{b})(x^2 + 3)$$

*Notice that in steps (1) – (3) you have written the polynomial so that we can see its binomial GCF. In step (4) we are really doing all of steps (1) – (5) from before.*

5. Check your answer.

Is $(3 + b)(x^2 + 3) = 3x^2 + 9 + bx^2 + 3b$?

Is $3(x^2 + 3) + b(x^2 + 3) = 3x^2 + 9 + bx^2 + 3b$?

Is $3x^2 + 9 + bx^2 + 3b = 3x^2 + 9 + bx^2 + 3b$? Yes.

## Sample Problems

1. Factor: $x(x^2 + y) + (-3)(x^2 + y)$

☐ a. Identify the terms of the polynomial. $x(x^2 + y)$ and ________

☐ b. Factor each term.
$x(x^2 + y) = x \cdot (x^2 + y)$
$(-3)(x^2 + y) =$ ________ $\cdot$ ________

☐ c. Find the GCF of the terms. GCF = ________

☐ d. Rewrite each term of the polynomial using the GCF.
$x(x^2 + y) =$ ________ $\cdot (x^2 + y)$
$(-3)(x^2 + y) =$ ________ $\cdot (x^2 + y)$

☐ e. Factor out the GCF. = (________)(________)

☐ f. Check your answer.

2. Factor: $x^2 + xy + 3x + 3y$

☐ a. Factor each term.
$x^2 = x \cdot x$
$xy = x \cdot y$
$3x =$ ___ $\cdot$ ___
$3y =$ ___ $\cdot$ ___

☐ b. Group terms with common factors.
$x^2 + xy + 3x + 3y$
$= (x \cdot x + x \cdot$ ___ $) + (3 \cdot$ ___ $+ 3 \cdot y)$

☐ c. Factor out the GCF in each grouping. = ___$(x + y)$ + ___$(x + y)$

☐ d. Factor out the binomial GCF of the polynomial. $= (x + y)$(________)

☐ e. Check your answer.

### *Answers to Sample Problems*

1.

*a.* $(-3)(x^2 + y)$

*b.* $-3$, $(x^2 + y)$ *(in either order)*

*c.* $x^2 + y$

*d.* $x$
$-3$

*e.* $x^2 + y$; $x + (-3)$ or $x - 3$ *(in either order)*

*f.* $(x^2 + y)[x + (-3)] =$
$(x^2 + y)(x) + (x^2 + y)(-3)$

2.

*a.* 3, $x$ *(in either order)*
3, $y$ *(in either order)*

*b.* $y$, $x$

*c.* $x$, 3

*d.* $x + 3$

*e.* $(x + y)(x + 3)$
$= x(x + 3) + y(x + 3)$
$= x^2 + 3x + xy + 3y$
$= x^2 + xy + 3x + 3y$

# HOMEWORK

## Homework Problems

Circle the homework problems assigned to you by the computer, then complete them below.

### Explain

### Greatest Common Factor

1. Circle the expressions below that are monomials.

   $x^2 + 2$ $\qquad$ $xy^2 + y^2x$

   $x^3yz^2$ $\qquad$ $x$

2. Circle the expressions below that are **not** monomials.

   $xzy^8$ $\qquad$ $\frac{4}{x}$

   $\frac{13x}{12}$ $\qquad$ $x^2z + zy^2$

3. Find the GCF of $12x^3y$ and $6xy^2$.
4. Find the GCF of $3xyz^3$, $z$, and $16yz$.
5. Factor: $x^2y + 6y^2$
6. Factor: $3x^2 + 9xy^3 - 12xy$
7. Factor: $4a^2 - 4b^2$
8. Factor: $3x^4yz + 3xyz + 9yz$
9. Factor: $6xy^3 - 4x^2y^2 + 2xy$
10. Factor: $16a^3b^2 + 20a^2b^4 - 8a^3b^3$
11. Factor: $17x^2y^2z^2 + 68x^{10}y^{32}z + 153x^9y^4z^{12}$
12. Factor: $x^2 + xy + xz$

### Factoring by Grouping

13. Find the binomial GCF: $(x^5 + y) + 6x^2(x^5 + y)$
14. Factor: $(x^5 + y) + 6x^2(x^5 + y)$
15. Find the binomial GCF:
    $(3x + y)(xy + yz) + x^2y(xy + yz) + z^3(xy + yz)$
16. Factor: $(3x + y)(xy + yz) + x^2y(xy + yz) + z^3(xy + yz)$
17. Factor: $a^3 - a^2b + ab^2 - b^3$
18. Factor: $3x^2 - 3xy + 3xy^3z^4 - 3y^4z^4$
19. Factor: $x^5y + zx + x^4y^2 + yz + x^4yz + z^2$
20. Factor: $15m^3 + 21m^2n + 10mn + 14n^2$
21. Factor: $x^2z + 3x^2 + y^2z + 3y^2$
22. Factor: $x^3 + x^2y + x^2z + 3x + 3z + 3y$
23. Factor: $3x + yz + xz + 3y$
24. Factor: $x^2 - 3x + 2$
    (Hint: rewrite the polynomial as $x^2 - x - 2x + 2$)

# APPLY

## Practice Problems

Here are some additional practice problems for you to try.

### Greatest Common Factor

1. Circle the expressions below that are monomials.

   $8m^3n$    $7y - 2y^2 + 14$    $x - y$

   $23$    $\frac{3}{z}$

2. Circle the expressions below that are monomials.

   $3x + 4x^2 - 7$    $17$    $5xyz^3$

   $y + z$    $\frac{1}{x}$

3. Find the GCF of $12a^3b$ and $16ab^4$.
4. Find the GCF of $18m^3n^5$ and $24m^4n^3$.
5. Find the GCF of $10xy^4$, and $15x^3y^2$.
6. Find the GCF of $9xy^2z^3$, $24x^5y^3z^6$, and $18x^3yz^4$.
7. Find the GCF of $6abc^4$, $12ac^3$, and $9a^5b^4c^2$.
8. Factor: $5a^3b + 10b$
9. Factor: $16mn^4 + 8m$
10. Factor: $6xy^2 + 12x$
11. Factor: $6x^4y^3 + 14xy$
12. Factor: $24mn - 16m^6n^2$
13. Factor: $8a^3b^2 - 10ab$
14. Factor: $24a^3b^4 + 42a^6b^5$
15. Factor: $36y^7z^8 - 45y^3z^5$
16. Factor: $25x^5y^7 + 35x^2y^4$
17. Factor: $4mn + 10mn^3 - 18m^4n$
18. Factor: $6xy + 9x^3y - 15xy^2$
19. Factor: $8a^3b^4 - 12ab + 20a^3b$
20. Factor: $15a^3b^4c^7 + 25a^5b^3c^2$
21. Factor: $32p^7q^3r^4 - 40p^5q^5r$
22. Factor: $24x^2y^5z^8 - 32x^4y^6z^4$
23. Factor: $9xy^2z^3 - 15x^3y^5z^4 + 21x^4y^2z^5$
24. Factor: $10h^4j^3k^6 + 25h^3j^2k - 40hj^5k^2$
25. Factor: $20a^3b^5c^2 + 12a^4b^2c^3 - 8a^2bc^3$
26. Factor: $20x^2y^4 + 10x^5y^3 - 18x^3y^4 + 12xy^3$
27. Factor: $6a^3b^5c^2 - 9a^4b^4c^3 + 18a^2b^3c^2 - 21a^6b^2c^3$
28. Factor: $18x^2y^4z^3 - 16x^5y^3z + 6x^4y^2z^3 - 10x^3y^4z^2$

### Factoring by Grouping

29. Factor: $x(z + 3) + y(z + 3)$
30. Factor: $a(b - 2) + c(b - 2)$
31. Factor: $a(3b - 4) + 9(3b - 4)$
32. Factor: $z(2w + 3) - 12(2w + 3)$
33. Factor: $8m(3n^3 - 4) + 17(3n^3 - 4)$
34. Factor: $12b(2c^4 + 5) - 23(2c^4 + 5)$
35. Factor: $7x(2x^2 + 3) - 11(2x^2 + 3)$
36. Factor: $a(3a - b) - b(3a - b)$
37. Factor: $m(5m + 2n) - 3n(5m + 2n)$
38. Factor: $y(2x + y) + x(2x + y)$
39. Factor: $xw + xz + yw + yz$
40. Factor: $mp - mq + np - nq$
41. Factor: $ac + ad - bc - bd$
42. Factor: $8a^2 + 4a + 10a + 5$

43. Factor: $4a^2 + 2a - 14a - 7$

44. Factor: $6x^2 - 2x + 12x - 4$

45. Factor: $12a^2 + 18a + 10ab + 15b$

46. Factor: $21m^2 - 14m + 24mn - 16n$

47. Factor: $15x^2 + 35x + 6xy + 14y$

48. Factor: $3u^2 + 6u + uv + 2v$

49. Factor: $8z^2 - 2z + 4zw - w$

50. Factor: $2x^2 + 4x - xy - 2y$

51. Factor: $12a^2 - 10b - 15ab + 8a$

52. Factor: $8m^2 + 21n + 12m + 14mn$

53. Factor: $18x^2 - 10y - 15xy + 12x$

54. Factor: $16uv^2 + 10vw + 25w + 40uv$

55. Factor: $12pr^2 - 16rs - 20s + 15pr$

56. Factor: $20ab^2 + 15bc - 6c - 8ab$

# EVALUATE

## Practice Test

Take this practice test to be sure that you are prepared for the final quiz in Evaluate.

1. Find the GCF of $6xz$, $3xy$, and $2x$.
2. Find the GCF of $16xyz$, $x^2y^2z^2$, and $4x^3y^2z$.
3. Factor: $3x^2y - 3xy^2$
4. Factor: $3xy^3 - 6xy^2 + 3x^3y^4$
5. Factor: $13(x^2 + 4) + 6y(x^2 + 4)$
6. Factor: $17x^2(3xyz + 4z) - 3yz(3xyz + 4z)$
7. Factor: $39rs - 13s + 9r - 3$
8. Factor: $12wz - 44z + 18w - 66$

# ANSWERS

## Homework

**1.** $x^3yz^2, x$ **3.** $6xy$ **5.** $y(x^2 + 6y)$

**7.** $4(a^2 - b^2) = 4(a + b)(a - b)$ **9.** $2xy(3y^2 - 2xy + 1)$

**11.** $17x^2y^2z(z + 4x^8y^{30} + 9x^7y^2z^{11})$ **13.** $x^5 + y$

**15.** $(xy + yz)$ **17.** $(a^2 + b^2)(a - b)$

**19.** $(x^4y + z)(x + y + z)$ **21.** $(x^2 + y^2)(z + 3)$

**23.** $(x + y)(3 + z)$

## Practice Problems

**1.** 23 and $8m^3n$ **3.** $4ab$ **5.** $5xy^2$ **7.** $3ac^2$

**9.** $8m(2n^4 + 1)$ **11.** $2xy(3x^3y^2 + 7)$ **13.** $2ab(4a^2b - 5)$

**15.** $9y^3z^5(4y^4z^3 - 5)$ **17.** $2mn(2 + 5n^2 - 9m^3)$

**19.** $4ab(2a^2b^3 - 3 + 5a^2)$ **21.** $8p^5q^3r(4p^2r^3 - 5q^2)$

**23.** $3xy^2z^3(3 - 5x^2y^3z + 7x^3z^2)$

**25.** $4a^2bc^2(5ab^4 + 3a^2bc - 2c)$

**27.** $3a^2b^2c^2(2ab^3 - 3a^2b^2c + 6b - 7a^4c)$

**29.** $(x + y)(z + 3)$ **31.** $(a + 9)(3b - 4)$

**33.** $(8m + 17)(3n^3 - 4)$ **35.** $(7x - 11)(2x^2 + 3)$

**37.** $(m - 3n)(5m + 2n)$ **39.** $(x + y)(w + z)$

**41.** $(a - b)(c + d)$ **43.** $(2a - 7)(2a + 1)$

**45.** $(6a + 5b)(2a + 3)$ **47.** $(5x + 2y)(3x + 7)$

**49.** $(2z + w)(4z - 1)$ **51.** $(4a - 5b)(3a + 2)$

**53.** $(6x - 5y)(3x + 2)$ **55.** $(3pr - 4s)(4r + 5)$

## Practice Test

**1.** GCF = $x$ **2.** GCF = $xyz$ **3.** $3x^2y - 3xy^2 = 3xy(x - y)$

**4.** $3xy^3 - 6xy^2 + 3x^3y^4 = 3xy^2(y - 2 + x^2y^2)$

**5.** $13(x^2 + 4) + 6y(x^2 + 4)$
$= (x^2 + 4)(13 + 6y)$

**6.** $17x^2(3xyz + 4z) - 3yz(3xyz + 4z)$
$= (3xyz + 4z)(17x^2 - 3yz)$

**7.** $39rs - 13s + 9r - 3$
$= (3r - 1)(13s + 3)$

**8.** $12wz - 44z + 18w - 66$
$= 2(3w - 11)(2z + 3)$

# LESSON 7.2 – FACTORING POLYNOMIALS II



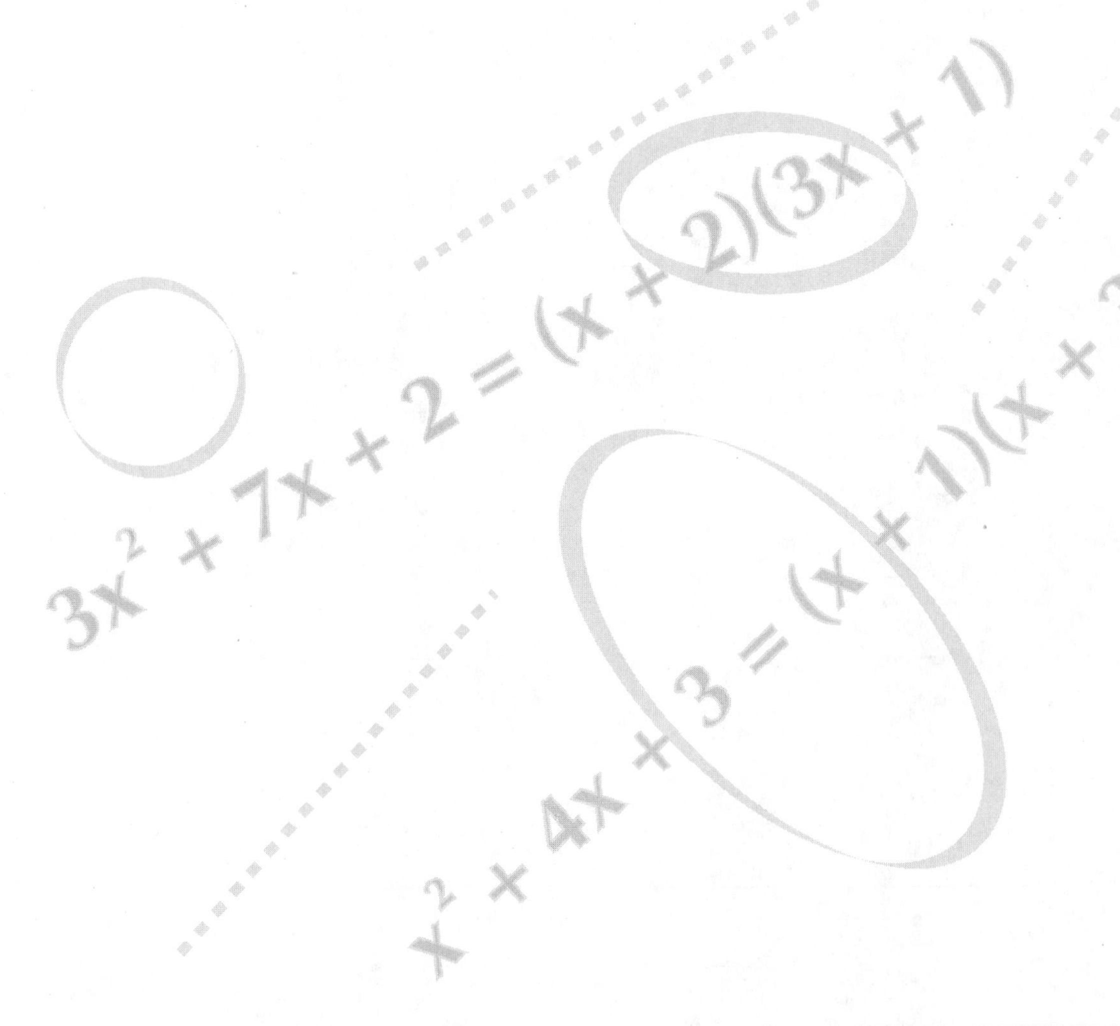

# OVERVIEW

***Here's what you'll learn in this lesson:***

***Trinomials I***

*a. Factoring trinomials of the form $x^2 + bx + c$; $x^2 + bxy + cy^2$*

***Trinomials II***

*a. Factoring trinomials of the form $ax^2 + bx + c$, $a \neq 1$, by trial-and-error*

*b. Factoring trinomials of the form $ax^2 + bx + c$, $a \neq 1$, by grouping*

*c. Solving quadratic equations by factoring*

You have already learned how to factor certain polynomials by finding the greatest common factor (GCF) and by grouping.

In this lesson, you will learn techniques for factoring trinomials. Then you will see how to use factoring to solve certain equations.

# EXPLAIN

## TRINOMIALS I

### Summary

#### Factoring Polynomials of the Form $x^2 + bx + c$

One way to factor a polynomial of the form $x^2 + bx + c$ as a product of binomials is to use the FOIL method, but work backwards. Here's an example.

The product of the first terms is $x^2$

$$x^2 - 3x - 4 = (x \quad)(x \quad)$$

The product of the last terms is $-4$

$$x^2 - 3x - 4 = (x \quad)(x \quad)$$

Try all the possible factorizations for which the product of the first terms is $x^2$ and the product of the last terms is $-4$. Since the product of the last terms is negative, one of the last terms is positive and the other is negative. Use the FOIL method to find factors whose "inner" and "outer" products add together to make $-3x$.

1. Make a chart of the possibilities for the binomial factors. These are shown in the table.

| possible factorizations |
|---|
| $(x + \mathbf{4})(x - \mathbf{1})$ |
| $(x - \mathbf{4})(x + \mathbf{1})$ |
| $(x + \mathbf{2})(x - \mathbf{2})$ |

2. Use the FOIL method to multiply the possible factorizations you listed in step (1). These are shown in the table.

| possible factorizations |
|---|
| $(x + \mathbf{4})(x - \mathbf{1}) = x^2 + 3x - 4$ |
| $(x - \mathbf{4})(x + \mathbf{1}) = x^2 - 3x - 4$ |
| $(x + \mathbf{2})(x - \mathbf{2}) = x^2 - 4$ |

3. Find the factorization that gives the original polynomial, $x^2 - 3x - 4$. In the second row you see that $x^2 - 3x - 4 = (x - 4)(x + 1)$.

   So the factorization is: $x^2 - 3x - 4 = (x - 4)(x + 1)$.

**Answers to Sample Problems**

b. $x^2 + 3x + 2$

c. $(x + 1)(x + 2)$ (in either order)

a. 6

3

## Sample Problems

1. Factor: $x^2 + 3x + 2$

☑ a. List all the possible factorizations where:

- the product of the first terms is $x^2$
- the product of the last terms is $+2$

Since the product of the last terms is positive, both of the last terms are positive or both are negative.

| possible factorizations |
|---|
| $(x + 1)(x + 2)$ |
| $(x - 1)(x - 2)$ |

☐ b. Multiply the possible factorizations. Identify the factorization that gives the middle term $+3x$.

| possible factorizations |
|---|
| $(x + 1)(x + 2)$ = ____________ |
| $(x - 1)(x - 2)$ = $x^2 - 3x + 2$ |

☐ c. Write the correct factorization. $x^2 + 3x + 2 =$________________

2. Factor: $x^2 - 7x + 12$

☐ a. List all the possible factorizations where:

- the product of the first terms is $x^2$
- the product of the last terms is $+12$

Since the product of the last terms is positive, both of the last terms are positive or both are negative.

| possible factorizations |
|---|
| $(x + 1)(x + 12)$ |
| $(x - 1)(x - 12)$ |
| $(x + 2)(x + 6)$ |
| $(x - 2)(x - \_\_)$ |
| $(x + \_\_)(x + 4)$ |
| $(x - 3)(x - 4)$ |

☐ b. Multiply the possible factorizations. Identify the factorization that gives the middle term $-7x$.

| possible factorizations |
|---|
| $(x+1)(x+12) = x^2+13x+12$ |
| $(x-1)(x-12) =$ ___________ |
| $(x+2)(x+6) =$ ___________ |
| $(x-2)(x-\_\_) = x^2-8x+12$ |
| $(x+\_\_)(x+4) = x^2+7x+12$ |
| $(x-3)(x-4) =$ ___________ |

☐ c. Write the correct factorization. $x^2-7x+12 =$ ___________

3. Factor: $x^2+x-2$

☐ a. List all the possible factorizations where:

- the product of the first terms is $x^2$
- the product of the last terms is $-2$

Since the product of the last terms is negative, one of the last terms is positive and the other is negative.

| possible factorizations |
|---|
| $(x+1)(x-2)$ |
| $(x-1)(\_\_\_\_)$ |

☐ b. Multiply the possible factorizations.
Identify the factorization that gives the middle term $+1x$.

| possible factorizations |
|---|
| $(x+1)(x-2) =$ ___________ |
| $(x-1)(\quad) =$ ___________ |

☐ c. Write the correct factorization. $x^2+x-2 =$ ___________.

***Answers to Sample Problems***

*b.* $x^2-13x+12$

$x^2+8x+12$

6

3

$x^2-7x+12$

*c.* $(x-3)(x-4)$ *(in either order)*

*a.* $x+2$

*b.* $x^2-x-2$

$x+2$, $x^2+x-2$

*c.* $(x-1)(x+2)$ *(in either order)*

**Answers to Sample Problems**

a. $x+2$

b. $x^2-x-2$

$x+2$, $x^2+x-2$

4. Factor: $x^2+2x-2$

☐ a. List all the possible factorizations where:

- the product of the first terms is $x^2$
- the product of the last terms is –2

Since the product of the last terms is negative, one of the last terms is positive and the other is negative.

| possible factorizations |
|---|
| $(x+1)(x-2)$ |
| $(x-1)($_____$)$ |

☐ b. Multiply the possible factorizations.
Identify the factorization that gives the middle term $+2x$.

| possible factorizations |
|---|
| $(x+1)(x-2)$ = ____________ |
| $(x-1)($_____$)$ = ____________ |

☑ c. Write the correct factorization. Neither of the possible factorizations gives the original polynomial, $x^2+2x-2$. So, $x^2+2x-2$ cannot be factored using integers.

# TRINOMIALS II

## Summary

### Factoring Polynomials of the Form $ax^2 + bx + c$ by Trial and Error

You have learned how to factor trinomials of the form $x^2 + bx + c$, where $b$ and $c$ are integers. Notice that the coefficient of $x^2$ is 1.

Now you will see how to factor trinomials of the form $ax^2 + bx + c$, where $a$, $b$, and $c$ are integers. Notice that the coefficient of $x^2$ can be an integer other than 1.

One way to factor a trinomial of the form $ax^2 + bx + c$ as a product of binomials is by trial and error. Here's an example.

Factor the trinomial $3x^2 - 14x - 5$ using trial and error. Notice that any factorization of this trinomial must look like this:

$$3x^2 - 14x - 5 = (\mathbf{?}x \quad \mathbf{?})(\mathbf{?}x \quad \mathbf{?})$$

The product of the $x$-terms must be $3x^2$ and the product of the constants must be $-5$. Since the product of the constants is negative, one of the constants is positive and the other is negative.

1. Make a chart of the possibilities for the $x$-terms in the binomial factors and possibilities for the constant terms in the binomial factors. These are shown in the table below.

| $x$-terms | constants |
|---|---|
| $3x$, $x$ | 1, −5 |
| $3x$, $x$ | 5, −1 |
| $3x$, $x$ | −1, 5 |
| $3x$, $x$ | −5, 1 |

2. Use the values from step (1) to list possible factorizations. These are shown in the table below.

| $x$-terms | constants | possible factorizations |
|---|---|---|
| $3x$, $x$ | 1, −5 | $(3x + 1)(x - 5)$ |
| $3x$, $x$ | 5, −1 | $(3x + 5)(x - 1)$ |
| $3x$, $x$ | −1, 5 | $(3x - 1)(x + 5)$ |
| $3x$, $x$ | −5, 1 | $(3x - 5)(x + 1)$ |

3. Use the FOIL method to do the multiplication of the possible factorizations you listed in step (2). These are shown in the table below.

| $x$-terms | constants | possible factorizations |
|---|---|---|
| $3x, x$ | $1, -5$ | $(3x + 1)(x - 5) = 3x^2 - 14x - 5$ |
| $3x, x$ | $5, -1$ | $(3x + 5)(x - 1) = 3x^2 + 2x - 5$ |
| $3x, x$ | $-1, 5$ | $(3x - 1)(x + 5) = 3x^2 + 14x - 5$ |
| $3x, x$ | $-5, 1$ | $(3x - 5)(x + 1) = 3x^2 - 2x - 5$ |

4. Find the factorization that equals the original polynomial, $3x^2 - 14x - 5$.
You can see that the shaded row is $3x^2 - 14x - 5$. So the factorization is:
$3x^2 - 14x - 5 = (3x + 1)(x - 5)$

Here's another example. Factor the trinomial $15x^2 - 16x + 4$ using trial and error. Notice that any factorization of this trinomial must look like this:

$15x^2 - 16x + 4 = (\mathbf{?}x \quad \mathbf{?})(\mathbf{?}x \quad \mathbf{?})$

The product of the $x$-terms must be $15x^2$ and the product of the constant terms must be $+4$. Since the product of the last terms is positive, both of the last terms are positive or both are negative.

1. Make a chart of the possibilities for the $x$-terms in the binomial factors and possibilities for the constant terms in the binomial factors. These are shown in the table below.

| $x$-terms | constants |
|---|---|
| $x, 15x$ | $1, 4$ |
| $x, 15x$ | $2, 2$ |
| $x, 15x$ | $4, 1$ |
| $x, 15x$ | $-1, -4$ |
| $x, 15x$ | $-2, -2$ |
| $x, 15x$ | $-4, -1$ |
| $3x, 5x$ | $1, 4$ |
| $3x, 5x$ | $2, 2$ |
| $3x, 5x$ | $4, 1$ |
| $3x, 5x$ | $-1, -4$ |
| $3x, 5x$ | $-2, -2$ |
| $3x, 5x$ | $-4, -1$ |

2. Use the values from step (1) to list possible factorizations. These are shown in the table that follows.

| $x$-terms | constants | possible factorizations |
|---|---|---|
| $x$, $15x$ | 1, 4 | $(x + 1)(15x + 4)$ |
| $x$, $15x$ | 2, 2 | $(x + 2)(15x + 2)$ |
| $x$, $15x$ | 4, 1 | $(x + 4)(15x + 1)$ |
| $x$, $15x$ | –1, –4 | $(x - 1)(15x - 4)$ |
| $x$, $15x$ | –2, –2 | $(x - 2)(15x - 2)$ |
| $x$, $15x$ | –4, –1 | $(x - 4)(15x - 1)$ |
| $3x$, $5x$ | 1, 4 | $(3x + 1)(5x + 4)$ |
| $3x$, $5x$ | 2, 2 | $(3x + 2)(5x + 2)$ |
| 3x, 5x | 4, 1 | $(3x + 4)(5x + 1)$ |
| $3x$, $5x$ | –1, –4 | $(3x - 1)(5x - 4)$ |
| $3x$, $5x$ | –2, –2 | $(3x - 2)(5x - 2)$ |
| $3x$, $5x$ | –4, –1 | $(3x - 4)(5x - 1)$ |

3. Use the FOIL method to do the multiplication of the possible factorizations you listed in step (2). These are shown in the table.

| $x$-terms | constants | possible factorizations |
|---|---|---|
| $x$, $15x$ | 1, 4 | $(x + 1)(15x + 4) = 15x^2 + 19x + 4$ |
| $x$, $15x$ | 2, 2 | $(x + 2)(15x + 2) = 15x^2 + 32x + 4$ |
| $x$, $15x$ | 4, 1 | $(x + 4)(15x + 1) = 15x^2 + 61x + 4$ |
| $x$, $15x$ | –1, –4 | $(x - 1)(15x - 4) = 15x^2 - 19x + 4$ |
| $x$, $15x$ | –2, –2 | $(x - 2)(15x - 2) = 15x^2 - 32x + 4$ |
| $x$, $15x$ | –4, –1 | $(x - 4)(15x - 1) = 15x^2 - 61x + 4$ |
| $3x$, $5x$ | 1, 4 | $(3x + 1)(5x + 4) = 15x^2 + 17x + 4$ |
| 3x, 5x | 2, 2 | $(3x + 2)(5x + 2) = 15x^2 + 16x + 4$ |
| $3x$, $5x$ | 4, 1 | $(3x + 4)(5x + 1) = 15x^2 + 23x + 4$ |
| $3x$, $5x$ | –1, –4 | $(3x - 1)(5x - 4) = 15x^2 - 17x + 4$ |
| $3x$, $5x$ | –2, –2 | $(3x - 2)(5x - 2) = 15x^2 - 16x + 4$ |
| $3x$, $5x$ | –4, –1 | $(3x - 4)(5x - 1) = 15x^2 - 23x + 4$ |

4. Find the factorization that equals the original polynomial, $15x^2 - 16x + 4$. You can see that the shaded row is $15x^2 - 16x + 4$. So the factorization is: $15x^2 - 16x + 4 = (3x - 2)(5x - 2)$

Here's another example. Factor the trinomial $3x^2 - 8x - 5$ using trial and error. Notice that any factorization of this trinomial must look like this:
$3x^2 - 8x - 5 =$ (**?**$x$ **?**)(**?**$x$ **?**)

The product of the $x$-terms must be $3x^2$ and the product of the constants must be –5. Since the product of the constants is negative, one of the constants is positive and the other is negative.

1. Make a chart of the possibilities for the $x$-terms in the binomial factors and possibilities for the constant terms in the binomial factors. These are shown in the table below.

| $x$-terms | constants |
|---|---|
| $3x$, $x$ | 1, –5 |
| $3x$, $x$ | 5, –1 |
| $3x$, $x$ | –1, 5 |
| $3x$, $x$ | –5, 1 |

2. Use the values from step (1) to list possible factorizations. These are shown in the table below.

| $x$-terms | constants | possible factorizations |
|---|---|---|
| $3x$, $x$ | 1, –5 | $(3x + 1)(x - 5)$ |
| $3x$, $x$ | 5, –1 | $(3x + 5)(x - 1)$ |
| $3x$, $x$ | –1, 5 | $(3x - 1)(x + 5)$ |
| $3x$, $x$ | –5, 1 | $(3x - 5)(x + 1)$ |

3. Use the FOIL method to do the multiplication of the possible factorizations you listed in step (2). These are shown in the table below.

| $x$-terms | constants | possible factorizations |
|---|---|---|
| $3x$, $x$ | 1, –5 | $(3x + 1)(x - 5) = 3x^2 - 14x - 5$ |
| $3x$, $x$ | 5, –1 | $(3x + 5)(x - 1) = 3x^2 + 2x - 5$ |
| $3x$, $x$ | –1, 5 | $(3x - 1)(x + 5) = 3x^2 + 14x - 5$ |
| $3x$, $x$ | –5, 1 | $(3x - 5)(x + 1) = 3x^2 - 2x - 5$ |

4. Find the factorization that equals the original polynomial, $3x^2 - 8x - 5$. You can see that no row is $3x^2 - 8x - 5$. So, $3x^2 - 8x - 5$ cannot be factored using integers.

## Factoring Polynomials of the Form $ax^2 + bx + c$ by Grouping

Another way to factor a trinomial of the form $ax^2 + bx + c$ is by grouping.

Remember how to multiply binomials using the FOIL method.

$$(x + 2)(3x + 1) = 3x^2 + x + 6x + 2$$
$$= 3x^2 + 7x + 2$$

To factor $3x^2 + 7x + 2$, we go the other way. We first write $3x^2 + 7x + 2$ using two $x$-terms, like this:

$$3x^2 + x + 6x + 2$$

Now, factor $3x^2 + x + 6x + 2$ by grouping:

1. Factor each term.

$$3x^2 = 3 \cdot x \cdot x$$
$$x = x$$
$$6x = 2 \cdot 3 \cdot x$$
$$2 = 2$$

2. Group terms with common factors. $= (3x^2 + x) + (6x + 2)$

3. Factor out the GCF in each grouping. $= x(3x + 1) + 2(3x + 1)$

4. Factor out the binomial GCF of the polynomial. $= (3x + 1)(x + 2)$

5. Check your answer. Is $(3x + 1)(x + 2) = 3x^2 + 7x + 2$ ?

   Is $3x^2 + 7x + 2 = 3x^2 + 7x + 2$ ? Yes.

In order to use grouping to factor this trinomial, you had to find two integers whose sum was 7 and whose product was 6.

To factor a trinomial of the form $ax^2 + bx + c$, you need to find two integers whose sum is $b$ and whose product is $ac$. Then you can split the $x$-term into two terms and factor by grouping.

For example, to factor $6x^2 + 7x + 2$ by grouping:

1. Make a chart of possible pairs of integers product is $6 \cdot 2 = 12$.

| possibilities | product | sum |
|---|---|---|
| 1, 12 | 12 | 13 |
| 2, 6 | 12 | 8 |
| 3, 4 | 12 | 7 |

*Notice that the chart doesn't include negative factors of 12. Can you see why not? Since the product of the two numbers has to be +12, if one factor is negative, both would have to be negative. But since the sum of the integers needs to be +7, a positive number, you know both factors can't be negative.*

2. Identify the numbers that work. Here, the last choice works since $3 + 4 = 7$ and $3 \cdot 4 = 12$.

3. Rewrite the trinomial. $6x^2 + 7x + 2 = 6x^2 + 3x + 4x + 2$

4. Group the terms. $= (6x^2 + 3x) + (4x + 2)$

5. Factor out the GCF in each grouping. $= 3x(2x + 1) + 2(2x + 1)$

6. Factor out the binomial GCF of the polynomial. $= (2x + 1)(3x + 2)$

7. Check your answer. Is $(2x + 1)(3x + 2) = 6x^2 + 7x + 2$?

   Is $6x^2 + 4x + 3x + 2 = 6x^2 + 7x + 2$? Yes.

## Solving Quadratic Equations of the Form $ax^2 + bx + c = 0$ by Factoring

You can use what you have learned about factoring to solve some quadratic equations.

A quadratic (or second-degree) equation in one variable is an equation that can be written in this form:

$$ax^2 + bx + c = 0$$

This is called standard form. Here, $a$, $b$, and $c$ are real numbers, and $a \neq 0$. The terms on the left side of the equation are in descending order by degree. The right side of the equation is zero.

If the left side of a quadratic equation in standard form can be factored, then you can solve the quadratic equation by factoring. To solve such an equation, you'll use a property called the Zero Product Property, which states the following: if P and Q are polynomials and if $P \cdot Q = 0$, then $P = 0$ or $Q = 0$ or both P and Q are 0.

Here's how to solve a quadratic equation in standard form when the left side can be factored:

1. Make sure the equation is in standard form.
2. Factor the left side.
3. Use the Zero Product Property. Set each factor equal to zero.
4. Finish solving for $x$.
5. Check your answer.

For example, to solve the equation $x^2 = 4x$:

1. Write the equation in standard form. $\quad x^2 - 4x = 0$
2. Factor the left side. $\quad x(x - 4) = 0$
3. Use the Zero Product Property to set each factor equal to zero. $\quad x = 0 \quad \text{or} \quad x - 4 = 0$
4. Finish solving for $x$. $\quad x = 0 \quad \text{or} \quad x = 4$
5. Check your answer.

| Check $x = 0$: | Check $x = 4$: |
|---|---|
| Is $0^2 = 4(0)$? | Is $4^2 = 4(4)$? |
| Is $0 = 0$ ? Yes. | Is $16 = 16$ ? Yes. |

So, both 0 and 4 are valid solutions of the equation $x^2 = 4x$.

## Sample Problems

1. Use trial and error to factor the polynomial $35x^2 + 73x + 6$.

   ☐ a. Write possible $x$-terms whose product is $35x^2$ and write possible constant terms whose product is 6.

   ☐ b. List the possible factorizations.

   ☐ c. Multiply the possible factorizations.

| $x$-terms | constants | possible factorizations |
|---|---|---|
| $x$, $35x$ | 1, 6 | $(x + 1)(35x + 6) = 35x^2 + 41x + 6$ |
| $x$, $35x$ | 2, __ | (_____)(_____) = __________ |
| $x$, $35x$ | 3, 2 | $(x + 3)(35x + 2) = 35x^2 + 107x + 6$ |
| $x$, $35x$ | 6, __ | (_____)(_____) = __________ |
| $x$, $35x$ | __, −6 | $(x - 1)(35x - 6) = 35x^2 - 41x + 6$ |
| $x$, $35x$ | −2, __ | (_____)(_____) = __________ |
| $x$, $35x$ | −3, −2 | $(x - 3)(35x - 2) = 35x^2 - 107x + 6$ |
| $x$, $35x$ | −6, __ | (_____)(_____) = __________ |
| $5x$, $7x$ | 1, 6 | $(5x + 1)(7x + 6) =$ __________ |
| $5x$, $7x$ | 2, __ | (_____)(_____) $= 35x^2 + 29x + 6$ |
| $5x$, $7x$ | 3, 2 | $(5x + 3)(7x + 2) = 35x^2 + 31x + 6$ |
| $5x$, $7x$ | 6, __ | $(5x + 6)(7x + 1) =$ __________ |
| $5x$, $7x$ | −1, __ | (_____)(_____) = __________ |
| $5x$, $7x$ | __, −3 | (_____)(_____) = __________ |
| $5x$, $7x$ | −3, −2 | $(5x - 3)(7x - 2) = 35x^2 - 31x + 6$ |
| $5x$, $7x$ | __, −1 | (_____)(_____) = __________ |

   ☐ d. Write the correct factorization. $35x^2 + 73x + 6 =$ __________
   ("in either order")

2. Use trial and error to factor $4x^2 - 4x - 15$.

   ☐ a. Write possible $x$-terms whose product is $4x^2$ and the possible constant terms whose product is −15.

   ☐ b. List the possible factorizations.

   ☐ c. Multiply the possible factorizations.

### *Answers to Sample Problems*

*a., b., c.*

*3, $(x + 2)(35x + 3) = 35x^2 + 73x + 6$*

*1, $(x + 6)(35x + 1) = 35x^2 + 211x + 6$*

*−1*

*−3, $(x - 2)(35x - 3) = 35x^2 - 73x + 6$*

*−1, $(x - 6)(35x - 1) = 35x^2 - 211x + 6$*

*$35x^2 + 37x + 6$*

*3, $(5x + 2)(7x + 3)$*

*1, $35x^2 + 47x + 6$*

*−6, $(5x - 1)(7x - 6) = 35x^2 - 37x + 6$*

*−2, $(5x - 2)(7x - 3) = 35x^2 - 29x + 6$*

*−6, $(5x - 6)(7x - 1) = 35x^2 - 47x + 6$*

*$(x + 2)(35x + 3)$*

***Answers to Sample Problems***

a., b., c.

$-1$, $x + 15$, $4x - 1$, $4x^2 + 59x - 15$

$-1$, $2x + 15$, $2x - 1$, $4x^2 + 28x - 15$

$15$, $x - 1$, $4x + 15$, $4x^2 + 11x - 15$

$15$, $2x - 1$, $2x + 15$, $4x^2 + 28x - 15$

$-3$, $x - 3$, $4x + 5$, $4x^2 - 7x - 15$

$-3$, $2x - 3$, $2x + 5$, $4x^2 + 4x - 15$

$-5$, $x - 5$, $4x + 3$, $4x^2 - 17x - 15$

$-5$, $2x - 5$, $2x + 3$, $4x^2 - 4x - 15$

$1$, $x - 15$, $4x + 1$, $4x^2 - 59x - 15$

$1$, $2x - 15$, $2x + 1$, $4x^2 - 28x - 15$

d. $2x - 5$, $2x + 3$ (in either order)

d. $6x^2 + 3x$, $8x + 4$

e. $3x$, $4$

f. $(3x + 4)$

| $x$-terms | constants | possible factorizations |
|---|---|---|
| $x$, $4x$ | 1, –15 | $(x + 1)(4x - 15) = 4x^2 - 11x - 15$ |
| $2x$, $2x$ | 1, –15 | $(2x + 1)(2x - 15) = 4x^2 - 28x - 15$ |
| $x$, $4x$ | 3, –5 | $(x + 3)(4x - 5) = 4x^2 + 7x - 15$ |
| $2x$, $2x$ | 3, –5 | $(2x + 3)(2x - 5) = 4x^2 - 4x - 15$ |
| $x$, $4x$ | 5, –3 | $(x + 5)(4x - 3) = 4x^2 + 17x - 15$ |
| $2x$, $2x$ | 5, –3 | $(2x + 5)(2x - 3) = 4x^2 + 4x - 15$ |
| $x$, $4x$ | 15, ___ | (_____)(_____) = ___________ |
| $2x$, $2x$ | 15, ___ | (_____)(_____) = ___________ |
| $x$, $4x$ | –1, ___ | (_____)(_____) = ___________ |
| $2x$, $2x$ | –1, ___ | (_____)(_____) = ___________ |
| $x$, $4x$ | ___, 5 | (_____)(_____) = ___________ |
| $2x$, $2x$ | ___, 5 | (_____)(_____) = ___________ |
| $x$, $4x$ | ___, 3 | (_____)(_____) = ___________ |
| $2x$, $2x$ | ___, 3 | (_____)(_____) = ___________ |
| $x$, $4x$ | –15, ___ | (_____)(_____) = ___________ |
| $2x$, $2x$ | –15, ___ | (_____)(_____) = ___________ |

☐ d. Write the correct factorization. $4x^2 - 4x - 15 =$ (________)(________)

3. Use grouping to factor $6x^2 + 11x + 4$.

☑ a. Make a chart of pairs of integers whose product is $6 \cdot 4 = 24$.

| possibilities | product | sum |
|---|---|---|
| 1, 24 | 24 | 25 |
| 2, 12 | 24 | 14 |
| 3, 8 | 24 | 11 |
| 4, 6 | 24 | 10 |

☑ b. Identify the two integers whose product is 24 and whose sum is 11. The two integers are 3 and 8.

☑ c. Rewrite the trinomial by splitting the $x$-term. $6x^2 + 11x + 4 = 6x^2 + 3x + 8x + 4$

☐ d. Group the terms. = (________) + (________)

☐ e. Factor out the GCF in each grouping. = ___ $(2x + 1)$ + ___ $(2x + 1)$

☐ f. Factor out the binomial GCF of the polynomial. $= (2x + 1)$(________)

- ☐ g. Check your answer.

4. Use grouping to factor $3x^2 - 4x - 15$.

   - ☐ a. Make a chart of pairs of integers whose product is $3 \cdot (-15) = -45$.

| possibilities | product | sum |
| --- | --- | --- |
| –1, 45 | –45 | 44 |
| –3, 15 | –45 | ___ |
| –5, ___ | –45 | ___ |
| 1, ___ | –45 | ___ |
| ___, –15 | –45 | ___ |
| ___, ___ | –45 | ___ |

   - ☐ b. Identify the two integers whose product is –45 and whose sum is –4. The two integers are ______ and ______.
   - ☐ c. Rewrite the trinomial by splitting the x-term. $3x^2 - 4x - 15 = 3x^2 +$ ___$x$ + ___$x - 15$
   - ☐ d. Group the terms. = (________) + (________)
   - ☐ e. Factor out the GCF in each grouping. = ___(______) + ___(______)
   - ☐ f. Factor out the binomial GCF of the polynomial. = (________)(________)
   - ☐ g. Check your answer.

5. Solve this quadratic equation for x by factoring: $8x^2 = 26x + 45$

   - ✔ a. Write the equation in standard form. $8x^2 - 26x - 45 = 0$
   - ☐ b. Factor the left side. (______)(______) = 0
   - ☐ c. Use the Zero Product Property. $4x + 5 =$ ___ or $2x - 9 =$ ___
   - ☐ d. Finish solving for x. $4x = -5$ or $2x = 9$

     $x =$ ___ or $x =$ ___
   - ☐ e. Check your answer.

### *Answers to Sample Problems*

*g.* $(2x + 1)(3x + 4)$
$= 2x(3x) + 2x(4) + 1(3x) + 1(4)$
$= 6x^2 + 8x + 3x + 4$
$= 6x^2 + 11x + 4$

*a.* 12
9, 4
–45, –44
3, –12
5, –9, –4

*b.* 5, –9 (in either order)

*c.* 5, –9 (in either order)

*d.* $(3x^2 - 9x) + (5x - 15)$
or $(3x^2 + 5x) - (9x + 15)$

*e.* $3x(x - 3) + 5(x - 3)$
or $x(3x + 5) - 3(3x + 5)$

*f.* $(x - 3)(3x + 5)$ in either order

*g.* $(x - 3)(3x + 5)$
$= 3x^2 + 5x - 9x - 15$
$= 3x^2 - 4x - 15$

*b.* $4x + 5$, $2x - 9$ (in either order)

*c.* 0, 0

*d.* $-\frac{5}{4}, \frac{9}{2}$

*e.* Is $8\left(-\frac{5}{4}\right)^2 = 26\left(-\frac{5}{4}\right) + 45$ ?

Is $8\left(\frac{25}{16}\right) = 26\left(-\frac{5}{4}\right) + 45$ ?

Is $\frac{25}{2} = -\frac{65}{2} + \frac{90}{2}$ ?

Is $\frac{25}{2} = \frac{25}{2}$ ? Yes.

Is $8\left(\frac{9}{2}\right)^2 = 26\left(\frac{9}{2}\right) + 45$ ?

Is $8\left(\frac{81}{4}\right) = 26\left(\frac{9}{2}\right) + 45$ ?

Is $162 = 117 + 45$ ?

Is $162 = 162$ ? Yes.

**Answers to Sample Problems**

# EXPLORE

## Sample Problems

On the computer you used overlapping circles to help find the GCF of a collection of monomials. You used a table to help factor polynomials. Below are some additional problems.

1. Use overlapping circles to find the GCF of $3x$ and $-9xy^3$.

☑ a. Factor each monomial. $3x = 3 \cdot x$

$-9xy^3 = -1 \cdot 3 \cdot 3 \cdot x \cdot y \cdot y \cdot y$

☑ b. Write the factorizations in the overlapping circles.

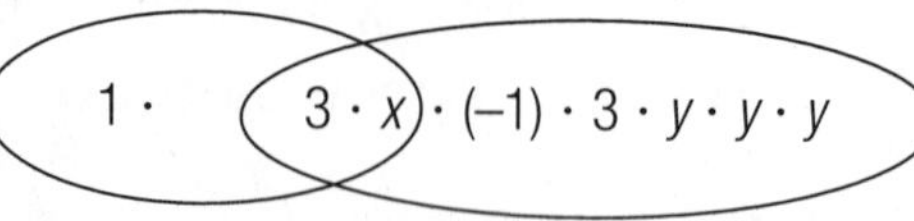


☐ c. Find the GCF from the overlapping circles. GCF = __________

*c. 3x*

2. Factor: $\frac{1}{5}x^2y - \frac{3}{5}xy$

☐ a. Factor each monomial. $\frac{1}{5}x^2y =$ __________

$\frac{3}{5}xy =$ __________

*a.* $\frac{1}{5} \cdot x \cdot x \cdot y$

$\frac{1}{5} \cdot 3 \cdot x \cdot y$

☐ b. Find the GCF of $\frac{1}{5}x^2y$ and $\frac{3}{5}xy$. GCF = __________

*b.* $\frac{1}{5}xy$

☐ c. Factor the polynomial $\frac{1}{5}x^2y - \frac{3}{5}xy$.

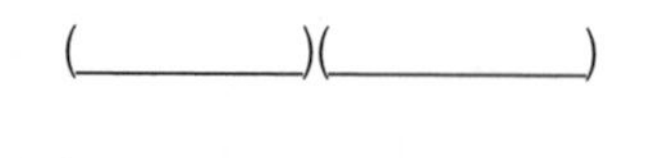

*c.* $\frac{1}{5}xy(x - 3)$

3. Find the GCF of the polynomials below.

   A: $22x^2z + 22yz$

   B: $11x^3 + 11xy$

   C: $2x^2 + 2y$

☐ a. Factor each polynomial.

$$22x^2z + 22yz = 2 \cdot 11 \cdot z(x^2 + y)$$

$$11x^3 + 11xy = \_\_\_\_\_\_\_\_\_\_$$

$$2x^2 + 2y = \_\_\_\_\_\_\_\_\_\_$$

☐ b. Finish writing the factorizations in the overlapping circles.

4. A trinomial with a missing constant term has been partially factored in the table below. Complete the table and write the polynomial and its factorization.

☐ a. What times $x$ gives $7x$? Use this to fill in box *a*.

☐ b. What times $3x$ gives $-9x$? Use this to fill in box b.

☐ c. Multiply boxes a and b. Use this to fill in box c.

☐ d. Write the polynomial and its factorization.

| | $x$ | b |
|---|---|---|
| $3x$ | $3x^2$ | $-9x$ |
| a | $7x$ | c |

(________) = (________) + (________)

***Answers to Sample Problems***

*a.* $11 \cdot x \cdot (x^2 + y)$

$2 \cdot (x^2 + y)$

*b.*

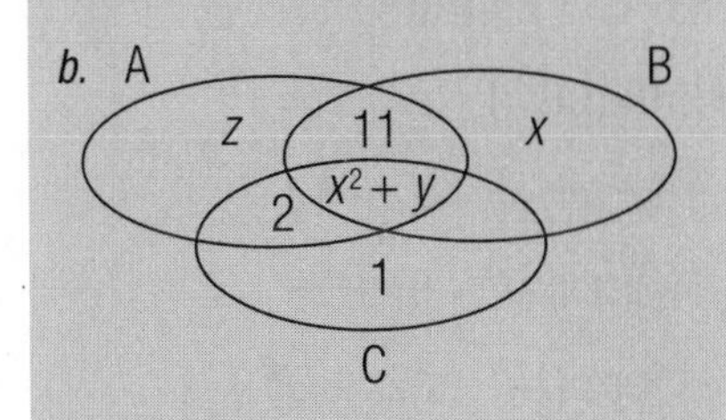


*a., b., c.*

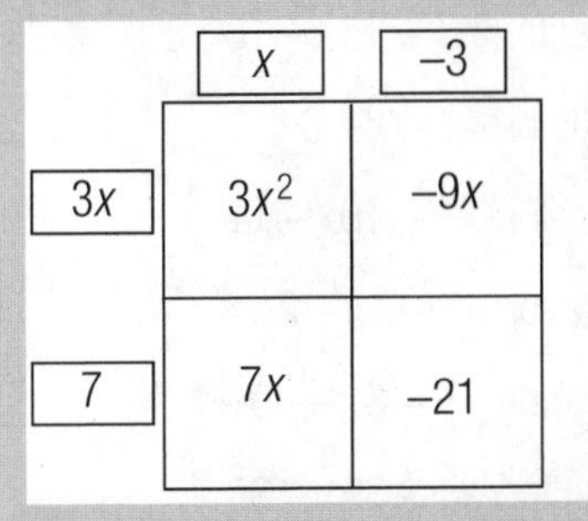

| | $x$ | $-3$ |
|---|---|---|
| $3x$ | $3x^2$ | $-9x$ |
| $7$ | $7x$ | $-21$ |

*d.*

$3x^2 - 2x - 21 = (3x + 7)(x - 3)$

# HOMEWORK

## Homework Problems

Circle the homework problems assigned to you by the computer, then complete them below.

### Explain

#### Trinomials I

1. Factor: $x^2 + 7x + 12$
2. Factor: $y^2 + 9y + 18$
3. Factor: $x^2 + 12x + 35$
4. Factor: $z^2 + 10z + 16$
5. Factor: $x^2 - 5x - 24$
6. Factor: $a^2 - 15a - 16$
7. Factor: $x^2 - x - 6$
8. Factor: $x^2 + 10x - 11$
9. Factor: $x^2 - 4x - 21$
10. Factor: $y^2 + 3y - 40$
11. Factor: $x^2 + 35x - 36$
12. Factor: $a^2 - 9a + 14$

#### Trinomials II

13. Factor: $2x^2 + 11x + 5$
14. Factor: $3x^2 + 13x + 4$
15. Factor: $4y^2 - 8y - 21$
16. Factor: $3z^2 - 17z + 20$
17. Factor: $15a^2 - 30a + 15$
18. Solve for $x$ by factoring: $6x^2 = 63 - 13x$
19. Solve for $x$ by factoring: $25x^2 + 5x = 2$
20. Factor: $4x^2 - 12x + 9$
21. Factor: $13x^2 + 37x + 22$
22. Factor: $x^2 - a^2$
23. Factor: $x^2 + 2xy + y^2$
24. Factor: $x^4 - 2ax^2 + a^2$

### Explore

25. Circle the monomial(s) below that might appear in the factorization of

    $3x^3y^2 + 2x^2y - 3xy$

    $3x^2y$ $\quad$ $2x^2y$ $\quad$ $xy$ $\quad$ $3x$

26. If the GCF of the terms of a polynomial is $4x^2y^3$, which of the monomials below could be terms in the polynomial?

    $4xy^3$ $\quad$ $8x^3y^4$ $\quad$ $4x^2y^3$ $\quad$ $4x^2$

27. Factor this polynomial using overlapping circles: $\frac{x^2y}{2} - \frac{2y}{4}$

28. A trinomial with a missing constant term has been partially factored in the table below. Complete the table and write the polynomial and its factorization.

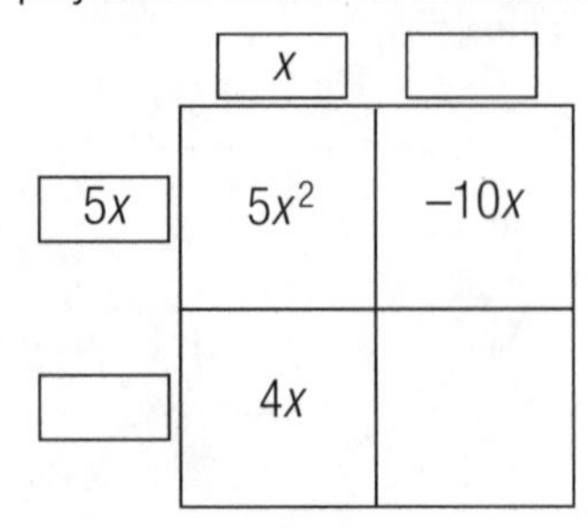


29. Complete the diagram below to find the GCF of the polynomials A, B, and C.

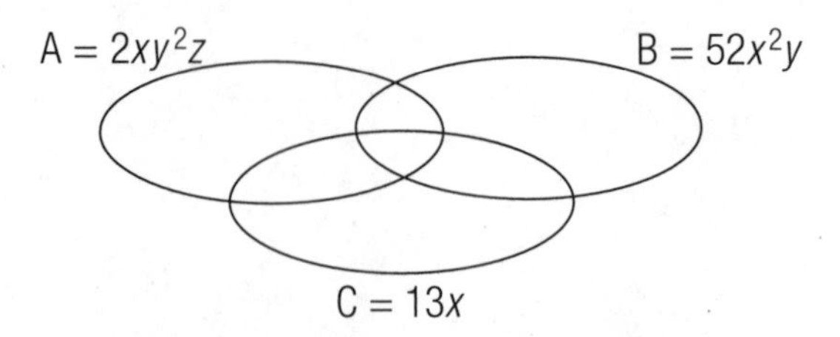


30. Factor this polynomial using overlapping circles:

    $\frac{1}{2}x^2y^2 + \frac{3}{2}x^3y^3 - 3x^2y$

# APPLY

## Practice Problems

Here are some additional practice problems for you to try.

### Trinomials I

1. Factor: $x^2 + 5x + 4$
2. Factor: $x^2 - 4x + 5$
3. Factor: $x^2 + 15x + 14$
4. Factor: $x^2 + 11x + 10$
5. Factor: $x^2 + 8x + 15$
6. Factor: $x^2 + 9x + 18$
7. Factor: $x^2 + 7x + 12$
8. Factor: $x^2 - 13x + 30$
9. Factor: $x^2 - 8x + 12$
10. Factor: $x^2 - 7x + 10$
11. Factor: $x^2 - 15x + 44$
12. Factor: $x^2 - 11x + 30$
13. Factor: $x^2 - 10x + 21$
14. Factor: $x^2 - 6x - 27$
15. Factor: $x^2 - 7x - 30$
16. Factor: $x^2 - 5x - 14$
17. Factor: $x^2 + 4x - 21$
18. Factor: $x^2 + 10x - 24$
19. Factor: $x^2 + 5x - 36$
20. Factor: $x^2 + 2x - 15$
21. Factor: $x^2 - 7x - 18$
22. Factor: $x^2 + 9x - 36$
23. Factor: $x^2 - 4x - 21$
24. Factor: $x^2 + 10x + 24$
25. Factor: $x^2 - 2x - 63$
26. Factor: $x^2 + 9x - 22$
27. Factor: $x^2 - 7x - 60$
28. Factor: $x^2 - 6x - 91$

### Trinomials II

29. Factor: $2x^2 + 7x + 5$
30. Factor: $2x^2 + 9x + 9$
31. Factor: $3x^2 - 19x - 14$
32. Factor: $2x^2 - 3x - 20$
33. Factor: $2x^2 - x - 28$
34. Factor: $3x^2 + 16x - 35$
35. Factor: $2x^2 + 5x - 12$
36. Factor: $2x^2 + 9x - 5$
37. Factor: $2x^2 + 13x + 15$
38. Factor: $2x^2 + 15x + 28$
39. Factor: $3x^2 + 11x + 6$
40. Factor: $12x^2 - 7x + 1$
41. Factor: $10x^2 - 9x + 2$
42. Factor: $6x^2 - 5x + 1$
43. Factor: $6x^2 - 11x - 10$
44. Factor: $9x^2 - 18x - 7$
45. Factor: $8x^2 - 2x - 3$
46. Factor: $6x^2 + 21x - 28$
47. Factor: $9x^2 - 3x - 20$
48. Factor: $4x^2 - 4x - 15$
49. Factor: $36x^2 + 13x + 1$
50. Factor: $30x^2 + 11x + 1$
51. Factor: $5x^2 + 14xy - 3y^2$
52. Factor: $4x^2 - 7xy - 2y^2$
53. Factor: $3x^2 - 5xy - 2y^2$
54. Factor: $6x^2 + xy - 12y^2$
55. Factor: $9x^2 - 3xy - 2y^2$
56. Factor: $4x^2 - 4xy - 3y^2$

# EVALUATE

## Practice Test

Take this practice test to be sure that you are prepared for the final quiz in Evaluate.

1. Factor: $x^2 - 10x + 24$

2. Circle the statement(s) below that are true.

   $x^2 + 2x - 1 = (x - 1)(x - 1)$

   $x^2 + 2x - 1 = (x + 2)(x - 1)$

   $x^2 + 2x - 1 = (x - 1)(x + 1)$

   $x^2 + 2x - 1 = (x + 1)(x + 1)$

   $x^2 + 2x - 1$ cannot be factored using integers

3. Factor: $t^2 - 16t - 17$

4. Factor: $r^2 + 10rt + 25t^2$

5. Factor: $5x^2 + 8x - 4$

6. Factor: $27v^2 - 57v + 28$

7. Factor: $4x^2 + 57x + 108$

8. Solve for $x$ by factoring: $7x^2 - 5x - 12 = 0$

9. The overlapping circles contain the factors of three monomials, A, B, and C.
   Circle the true statements below.

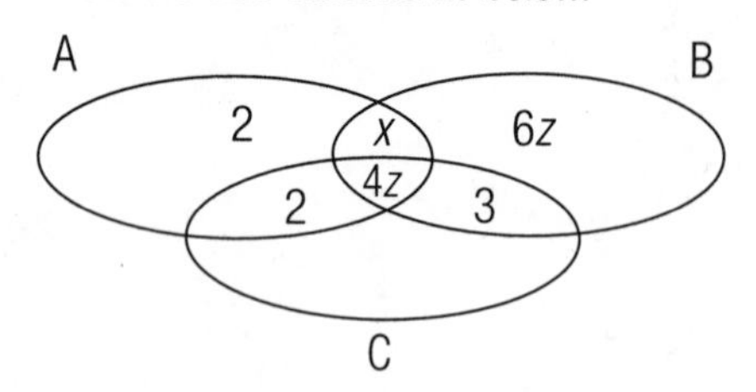


   Two factors of C are $z$ and 2.

   $B = 72xz$

   The GCF of A and B is $x$.

   The GCF of A, B, and C is $4z$.

10. The overlapping circles contain the factors of two binomials, A and B. Their GCF is $(3u + 4v)$. What are A and B?

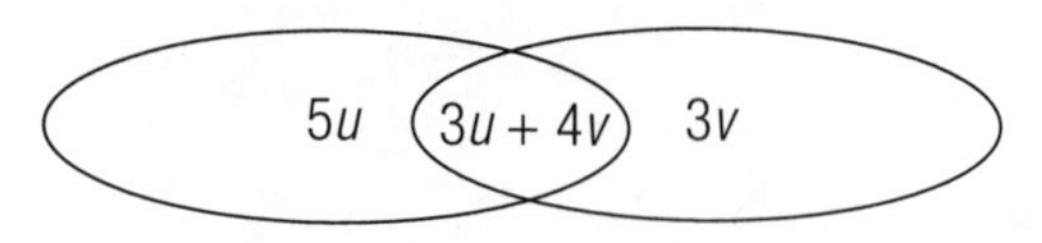


11. The polynomial $14xy + 21y - 6x^2 - 9x$ can be grouped as two binomials: $(14xy - 6x^2) + (21y - 9x)$. Find the GCF of the two binomials by factoring the polynomial using the overlapping circles below.

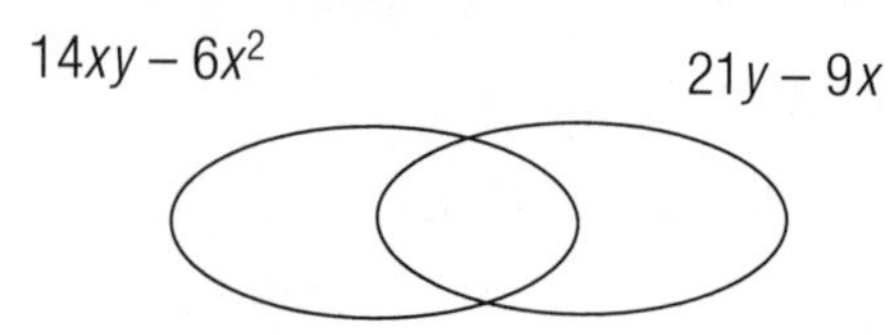


12. Finish factoring the trinomial $6x^2 - 7xy - 3y^2$ using the table below.

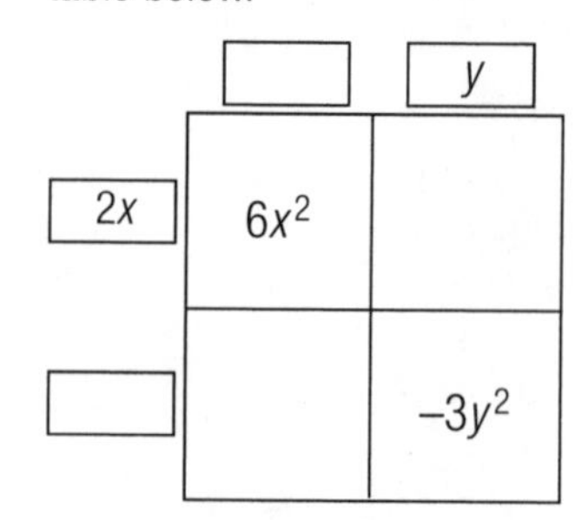

# ANSWERS

## Homework

**1.** $(x+4)(x+3)$ **3.** $(x+5)(x+7)$ **5.** $(x+3)(x-8)$

**7.** $(x-3)(x+2)$ **9.** $(x-7)(x+3)$

**11.** $(x+36)(x-1)$ **13.** $(x+5)(2x+1)$ **15.** $(2y+3)(2y-7)$

**17.** $15(a-1)(a-1)$ **19.** $x=\frac{1}{5}$ or $x=-\frac{2}{5}$

**21.** $(13x+11)(x+2)$ **23.** $(x+y)(x+y)$ **25.** $xy$

**27.** $\frac{y}{2}(x^2-1)$ or $\frac{1}{2}y(x^2-1)$

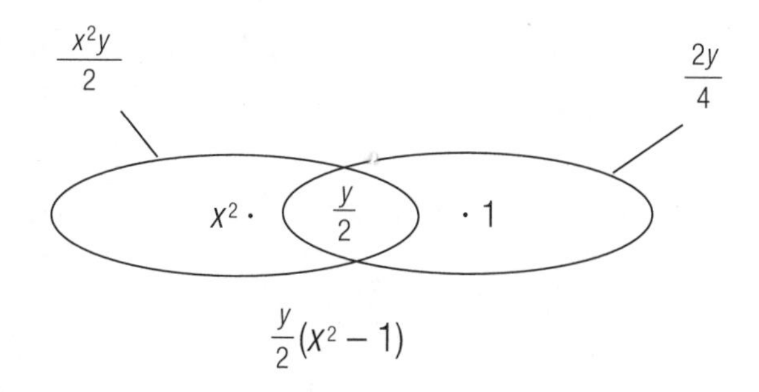


**29.** GCF = $x$

$A = 2xy^2z$ $B = 52x^2y$

$yz$ $2y$ $2x$

$x$ $13$

$C = 13x$

## Practice Problems

**1.** $(x+1)(x+4)$ **3.** $(x+1)(x+14)$ **5.** $(x+3)(x+5)$

**7.** $(x+4)(x+3)$ **9.** $(x-2)(x-6)$ **11.** $(x-11)(x-4)$

**13.** $(x-7)(x-3)$ **15.** $(x+3)(x-10)$ **17.** $(x+7)(x-3)$

**19.** $(x+9)(x-4)$ **21.** $(x-9)(x+2)$ **23.** $(x-7)(x+3)$

**25.** $(x-9)(x+7)$ **27.** $(x-12)(x+5)$ **29.** $(2x+5)(x+1)$

**31.** $(3x+2)(x-7)$ **33.** $(2x+7)(x-4)$ **35.** $(2x-3)(x+4)$

**37.** $(2x+3)(x+5)$ **39.** $(3x+2)(x+3)$ **41.** $(5x-2)(2x-1)$

**43.** $(3x+2)(2x-5)$ **45.** $(2x+1)(4x-3)$

**47.** $(3x-5)(3x+4)$ **49.** $(4x+1)(9x+1)$

**51.** $(5x-y)(x+3y)$ **53.** $(3x+y)(x-2y)$

**55.** $(3x-2y)(3x+y)$

## Practice Test

**1.** $x^2-10x+24=(x-4)(x-6)$

**2.** The polynomial $x^2+2x-1$ cannot be factored using integers.

**3.** $t^2-16t-17=(t+1)(t-17)$

**4.** $r^2+10rt+25t^2=(r+5t)(r+5t)$

**5.** $5x^2+8x-4=(x+2)(5x-2)$

**6.** $27v^2-57v+28=(9v-7)(3v-4)$

**7.** $4x^2+57x+108=(x+12)(4x+9)$

**8.** $x=\frac{12}{7}$ or $x=-1$

**9.** The two true statements are:

- Two factors of C are $z$ and 2.
- The GCF of A, B, and C is $4z$.

**10.** $15u^2+20uv$

$9uv+12v^2$

**11.** $7y-3x$

**12.**

| | [3] $x$ | [ $y$ ] |
|---|---|---|
| [ $2x$ ] | $6x^2$ | [2] $xy$ |
| [−3] $y$ | [ ] $xy$ | $-3y^2$ |

# LESSON 7.3 – FACTORING BY PATTERNS

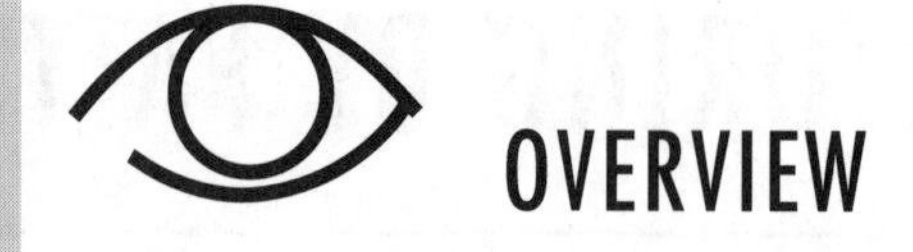

# OVERVIEW

***Here's what you'll learn in this lesson:***

***Recognizing Patterns***

*a. Factoring a perfect square trinomial*

*b. Factoring a difference of two squares*

*c. Factoring a sum and difference of two cubes*

*d. Factoring using a combination of methods*

A shortcut can make you more efficient by reducing the amount of time it takes to accomplish a task. It's always nice when you discover a shortcut: for taking notes in class, for programming a VCR, or for getting to a friend's house.

There are shortcuts that you can use in algebra to help you solve problems. For instance, when factoring polynomials, there are patterns you can look for that will help you factor the polynomials more quickly and accurately.

In this lesson you will learn how to recognize patterns for factoring polynomials.

# EXPLAIN

## RECOGNIZING PATTERNS

### Summary

Factoring by patterns means recognizing that polynomials having a certain form will always factor in a specific way. Perfect square trinomials, differences of two squares, and differences and sums of two cubes can all be factored using patterns. These patterns are described below.

*The patterns are just a shortcut. You can always factor using algebra tiles or trial and error.*

### Perfect Square Trinomials

One type of polynomial that's easy to factor using a pattern is a perfect square trinomial.

A perfect square trinomial is a polynomial that can be written so that it:

- has three terms
- has a first term that is a perfect square: $a^2$
- has a third term that is a perfect square: $b^2$
- has a second term that is twice the product of $a$ and $b$: $2ba$

For example, the polynomials below are perfect square trinomials:

$x^2 + 6x + 9$ $\quad$ $w^2 - 12w + 36$

$x^2 - 2xy + y^2$ $\quad$ $4y^4 + 24xy^2 + 36x^2$

The patterns for factoring perfect square trinomials are:

$$\triangle^2 + 2\square\triangle + \square^2 = (\triangle + \square)^2$$

$$a^2 + 2ba + b^2 = (a + b)^2$$

$$\triangle^2 - 2\square\triangle + \square^2 = (\triangle - \square)^2$$

$$a^2 - 2ba + b^2 = (a - b)^2$$

*Notice that the only difference between these two patterns is the sign of the middle term.*

For example, to factor $x^2 + 2x + 1$:

1. Decide which pattern to use. $\quad a^2 + 2ba + b^2 = (a + b)^2$
2. Substitute $x$ for $a$ and 1 for $b$. $\quad x^2 + 2(1)(x) + 1^2 = (x + 1)^2$

So, $x^2 + 2x + 1 = (x + 1)^2$.

As another example, to factor $x^2 - 4x + 4$:

1. Decide which pattern to use. $a^2 - 2ba + b^2 = (a - b)^2$
2. Substitute $x$ for $a$ and 2 for $b$. $x^2 - 2(2)(x) + 2^2 = (x - 2)^2$

So, $x^2 - 4x + 4 = (x - 2)^2$.

## Difference of Two Squares

Another type of polynomial that can be factored using a pattern is a difference of two squares.

A difference of two squares is a polynomial that can be written so that it:

- has two terms
- has a first term that is a perfect square: $a^2$
- has a second term that is a perfect square: $b^2$
- has a minus sign between the terms

*Why don't these polynomials have a middle term? Try multiplying $(a + b)(a - b)$ using the FOIL method and see what happens.*

*You can only use this pattern for factoring a **difference** of two squares. A **sum** of two squares can't be factored using integers.*

For example, the polynomials below are differences of two squares.

| | |
|---|---|
| $x^2 - 25$ | $y^2 - 100$ |
| $w^4 - x^2$ | $9x^2 - 81z^6$ |

The pattern for factoring a difference of two squares is:

$$\triangle^2 - \square^2 = (\triangle + \square)(\triangle - \square)$$
$$a^2 - b^2 = (a + b)(a - b)$$

For example, to factor $9y^2 - 25$:

1. Use this pattern. $a^2 - b^2 = (a + b)(a - b)$
2. Substitute $3y$ for $a$ and 5 for $b$. $(3y)^2 - 5^2 = (3y + 5)(3y - 5)$

So, $9y^2 - 25 = (3y + 5)(3y - 5)$.

## Differences or Sums of Two Cubes

You can also use patterns to factor a difference of two cubes or a sum of two cubes.These are polynomials that:

- have two terms
- have a first term that is a perfect cube: $a^3$
- have a second term that is a perfect cube: $b^3$

For example, the polynomials below are differences or sums of two cubes.

| | |
|---|---|
| $x^3 - 27$ | $64y^3 + 1$ |
| $x^3 + y^3$ | $8w^3 - 125x^6$ |

The patterns that can be used to factor a difference or sum of two cubes are:

$$\triangle^3 - \square^3 = (\triangle - \square)(\triangle^2 + \triangle\square + \square^2)$$
$$a^3 - b^3 = (a - b)(a^2 + ab + b^2)$$

$$\triangle^3 + \square^3 = (\triangle + \square)(\triangle^2 - \triangle\square + \square^2)$$
$$a^3 + b^3 = (a + b)(a^2 - ab + b^2)$$

For example, to factor $x^3 - 8$:

1. Decide which pattern to use. $a^3 - b^3 = (a - b)(a^2 + ab + b^2)$
2. Substitute $x$ for $a$ and 2 for $b$. $x^3 - 2^3 = (x - 2)(x^2 + 2x + 2^2)$

   $= (x - 2)(x^2 + 2x + 4)$

So, $x^3 - 8 = (x - 2)(x^2 + 2x + 4)$.

As another example, to factor $27y^3 + z^6$:

1. Decide which pattern to use. $a^3 + b^3 = (a + b)(a^2 - ab + b^2)$
2. Substitute $3y$ for $a$ and $z^2$ for $b$. $(3y)^3 + (z^2)^3 = (3y + z^2)[(3y)^2 - 3yz^2 + (z^2)^2]$

   $= (3y + z^2)(9y^2 - 3yz^2 + z^4)$

So, $27y^3 + z^6 = (3y + z^2)(9y^2 - 3yz^2 + z^4)$.

## Combining Patterns

When a polynomial doesn't look like it fits into one of the patterns you've seen, don't give up. Try to factor using another method first, then see if one of the polynomials in the factorization fits a pattern you have learned. For example, sometimes you need to factor out the greatest common factor of the terms before the polynomial will fit one of the patterns.

For example, to factor $2x^2y - 28xy + 98y$:

1. Factor out the GCF of the terms. $2y(x^2 - 14x + 49)$
2. Decide which pattern to use. $a^2 - 2ba + b^2 = (a - b)^2$
3. Substitute $x$ for $a$ and 7 for $b$. $x^2 - 2(7)(x) + 7^2 = (x - 7)^2$

So, $2x^2y - 28xy + 98y = 2y(x - 7)^2$.

Whatever factoring technique you use, look at the final product to be sure it cannot be factored any further.

For example, to factor $x^4 - y^4$:

1. Decide which pattern to use. $a^2 - b^2 = (a + b)(a - b)$
2. Substitute $x^2$ for $a$ and $y^2$ for $b$. $(x^2)^2 - (y^2)^2 = (x^2 + y^2)(x^2 - y^2)$

3. Determine if any factor can be factored further. — $x^2 - y^2$ can be factored further

4. Decide which pattern to use. — $a^2 - b^2 = (a + b)(a - b)$

5. Substitute $x$ for $a$ and $y$ for $b$. — $x^2 - y^2 = (x + y)(x - y)$

So, $x^4 - y^4 = (x^2 + y^2)(x^2 - y^2) = (x^2 + y^2)(x + y)(x - y)$.

## Sample Problems

1. Factor: $w^2 - 10w + 25$

☑ a. Decide which pattern to use.

__ $a^2 + 2ba + b^2 = (a + b)^2$

✓ $a^2 - 2ba + b^2 = (a - b)^2$

__ $a^2 - b^2 = (a + b)(a - b)$

__ $a^3 - b^3 = (a - b)(a^2 + ab + b^2)$

__ $a^3 + b^3 = (a + b)(a^2 - ab + b^2)$

☑ b. Substitute $w$ for $a$ and 5 for $b$.

$a^2 - 2ba + b^2 = (a - b)^2$

$w^2 - 2(5)(w) + 5^2 = (w - 5)^2$

2. Factor: $y^2 - 9$

☑ a. Decide which pattern to use. — $a^2 - b^2 = (a + b)(a - b)$

☐ b. Substitute $y$ for $a$ and 3 for $b$. — $y^2 - 9 = (\_\_\_\_\_\_)(\_\_\_\_\_\_)$

3. Factor: $27x^3 - 8$

☐ a. Decide which pattern to use. — $\_\_\_\_\_\_ = (\_\_\_\_\_\_)(\_\_\_\_\_\_\_\_\_\_\_\_)$

☐ b. Substitute $3x$ for $a$ and 2 for $b$. — $27x^3 - 8 = (\_\_\_\_\_\_)(\_\_\_\_\_\_\_\_\_\_\_\_)$

4. Factor: $12wx^2 - 27wy^2$

☐ a. Factor out the GCF of the terms. — $12wx^2 - 27wy^2 = 3w(\_\_\_\_\_\_\_\_\_\_)$

☐ b. Decide which pattern to use. — $\_\_\_\_\_\_\_\_\_\_ = (\_\_\_\_\_\_\_)(\_\_\_\_\_\_\_)$

☐ c. Substitute for $a$ and $b$. — $3w(\_\_\_\_\_\_) = 3w(\_\_\_\_\_\_)(\_\_\_\_\_)$

***Answers to Sample Problems***

2. *b.* $(y + 3)(y - 3)$

3. *a.* $a^3 - b^3$, $(a - b)(a^2 + ab + b^2)$

*b.* $(3x - 2)(9x^2 + 6x + 4)$

4. *a.* $4x^2 - 9y^2$

*b.* $a^2 - b^2$, $(a + b)(a - b)$

*c.* $4x^2 - 9y^2$, $2x + 3y$, $2x - 3y$

# HOMEWORK

## Homework Problems

Circle the homework problems assigned to you by the computer, then complete them below.

### Explain

### Recognizing Patterns

Factor the polynomials in problems 1 through 12.

1. $x^2 + 14x + 49$
2. $w^2 - 16$
3. $x^3 + 125$
4. $25y^2 - 30y + 9$
5. $9xy^2 - x$
6. $64y^3 - 27w^9$
7. $x^2 + 8w^2x + 16w^4$
8. $2x^6 - 72y^2$
9. $49y^2 - 28xy + 4x^2$
10. $x^3y^2 + 8y^2$
11. $2x^3 + 12x^2 + 18x$
12. $y^6 - 16y^2$

# APPLY

## Practice Problems

Here are some additional practice problems for you to try.

### Recognizing Patterns

1. Factor: $a^2 + 18a + 81$
2. Factor: $y^2 + 14y + 49$
3. Factor: $9x^2 + 42x + 49$
4. Factor: $25m^2 + 30m + 9$
5. Factor: $4a^2 + 20a + 25$
6. Factor: $b^2 - 16b + 64$
7. Factor: $z^2 - 22z + 121$
8. Factor: $y^2 - 18y + 81$
9. Factor: $16a^2 - 40a + 25$
10. Factor: $4c^2 + 28c + 49$
11. Factor: $9x^2 - 12x + 4$
12. Factor: $m^2 - 144$
13. Factor: $x^2 - 36$
14. Factor: $9m^2 - 81n^2$
15. Factor: $25a^2 - 625b^2$
16. Factor: $16x^2 - 64y^2$
17. Factor: $a^3 - 216$
18. Factor: $m^3 - 1000$
19. Factor: $x^3 - 125$
20. Factor: $8b^3 - 125$
21. Factor: $27z^3 - 343$
22. Factor: $64a^3 - 216$
23 Factor: $c^3 + 64$
24. Factor: $p^3 + 512$
25. Factor: $y^3 + 27$
26. Factor: $3a^3 + 42a^2b + 147b^2$
27. Factor: $50m^3n - 128mn^3$
28. Factor: $5x^3 - 20xy^2$

# EVALUATE

## Practice Test

Take this practice test to be sure that you are prepared for the final quiz in Evaluate.

1. Circle the expressions below that are perfect square trinomials.

   $9x^2 + 12x + 4$

   $0.25x^2 + 8x + 64$

   $25x^2 - 9$

   $9x^2 + 20x + 4$

   $x^2 - 2x + 1$

   $x^2 - 7x + 6$

2. Factor the polynomials below.

   a. $x^2 - 10x + 25$

   b. $49y^2 + 28y + 4$

   c. $16x^2 - 1$

   d. $9y^2 - 36$

3. Circle the polynomials below that **cannot** be factored any further using integers.

   $x^2 - 1000$

   $4y^2 - 4y + 1$

   $3x^2 - 27x + 9$

   $9m^2 - 24mn - 16n^2$

   $12x^3 - 8xy + 2y$

4. Factor: $12x^3 - 60x^2 + 75x$

5. Circle the expressions below that are perfect square trinomials.

   $36x^2 - 1$

   $4x^2 - 2x - 56$

   $x^2 + 8x + 16$

   $4x^2 - 12x + 9$

   $x^2 - 16x + 4$

6. Factor the polynomials below.

   a. $4x^2 - 24x + 36$

   b. $64z^2 + 16z + 1$

   c. $4w^2 - 49$

   d. $9m^2 - n^2$

7. Factor the polynomials below.

   a. $x^3 + 1000$

   b. $216y^3 - 1$

   c. $343x^3 + 8y^3$

8. Factor: $27w^3 + 90w^2 + 75w$

# ANSWERS

## Homework

**1.** $(x + 7)^2$ **3.** $(x + 5)(x^2 - 5x + 25)$ **5.** $x(3y - 1)(3y + 1)$
**7.** $(x + 4w^2)^2$ **9.** $(7y - 2x)^2$ **11.** $2x(x + 3)^2$

## Practice Problems

**1.** $(a + 9)(a + 9)$ or $(a + 9)^2$

**3.** $(3x + 7)(3x + 7)$ or $(3x + 7)^2$

**5.** $(2a + 5)(2a + 5)$ or $(2a + 5)^2$

**7.** $(z - 11)(z - 11)$ or $(z - 11)^2$

**9.** $(4a - 5)(4a - 5)$ or $(4a - 5)^2$

**11.** $(3x - 2)(3x - 2)$ or $(3x - 2)^2$

**13.** $(x + 6)(x - 6)$

**15.** $25(a + 5b)(a - 5b)$

**17.** $(a - 6)(a^2 + 6a + 36)$

**19.** $(x - 5)(x^2 + 5x + 25)$

**21.** $(3z - 7)(9z^2 + 21z + 49)$

**23.** $(c + 4)(c^2 - 4c + 16)$

**25.** $(y + 3)(y^2 - 3y + 9)$

**27.** $2mn(5m + 8n)(5m - 8n)$

## Practice Test

**1.** $9x^2 + 12x + 4$
$0.25x^2 + 8x + 64$
$x^2 - 2x + 1$

**2a.** $x^2 - 10x + 25 = (x - 5)(x - 5)$

**b.** $49y^2 + 28y + 4 = (7y + 2)(7y + 2)$

**c.** $16x^2 - 1 = (4x + 1)(4x - 1)$

**d.** $9y^2 - 36 = (3y + 6)(3y - 6)$

**3.** $x^2 - 1000$ cannot be factored any further using integers.
$9m^2 - 24mn - 16n^2$ cannot be factored any further using integers.

**4.** $12x^3 - 60x^2 + 75x = 3x(2x - 5)(2x - 5)$

**5.** $x^2 + 8x + 16$, $4x^2 - 12x + 9$

**6a.** $4x^2 - 24x + 36 = (2x - 6)(2x - 6)$

**b.** $64z^2 + 16z + 1 = (8z + 1)(8z + 1)$

**c.** $4w^2 - 49 = (2w + 7)(2w - 7)$

**d.** $9m^2 - n^2 = (3m + n)(3m - n)$

**7a.** $x^3 + 1000 = (x + 10)(x^2 - 10x + 100)$

**b.** $216y^3 - 1 = (6y - 1)(36y^2 + 6y + 1)$

**c.** $343x^3 + 8y^3 = (7x + 2y)(49x^2 - 14xy + 4y^2)$

**8.** $27w^3 + 90w^2 + 75w = 3w(3w + 5)(3w + 5)$

# TOPIC 7 CUMULATIVE ACTIVITIES

## CUMULATIVE REVIEW PROBLEMS

These problems combine all of the material you have covered so far in this course. You may want to test your understanding of this material before you move on to the next topic. Or you may wish to do these problems to review for a test.

1. Find: $(a^5 - 9a^3 + 5a^2 + 14a - 35) \div (a^2 - 7)$
2. Find the slope of the line perpendicular to the line through the points (–3, 7) and (9, –5).
3. Simplify this expression: $11x^2 + 6y + 2 - 4x^2 - y$
4. Graph the inequality $2x + 3y \leq 6$.
5. Alfredo needs to make 250 ml of a 27% alcohol solution using a 15% solution and a 40% solution. How much of each should he use?
6. The point (–2, –3) lies on a line with slope 2. Graph this line by finding another point that lies on the line.
7. Factor: $x^2 - 6x + 9$
8. Find the GCF of $6x^2y^2$ and $8xy^4$.
9. Circle the true statements below.

   $5(7 + 3) = 5(10)$

   $|9 - 20| = -11$

   The fraction $\frac{4}{6}$ is in lowest terms.

   The LCM of 45 and 75 is 225.

   $\frac{9}{11} \cdot \frac{22}{3} = 6$

10. Solve for $x$: $3(x + 1) = x + 2\left(x + \frac{3}{2}\right)$
11. Graph the system of inequalities below to find its solution.

    $3x - 2y \leq 7$
    $4x + y > 3$

12. Find the slope of the line parallel to the line through the points (7, –1) and (2, 8).
13. Factor: $a^4 + 4a^2 + 4$
14. Solve this system:

    $3x - y = 23$
    $2x + y = 22$

15. Solve $5 \leq 3x - 13 < 17$ for $x$, then graph the solution on the number line below.

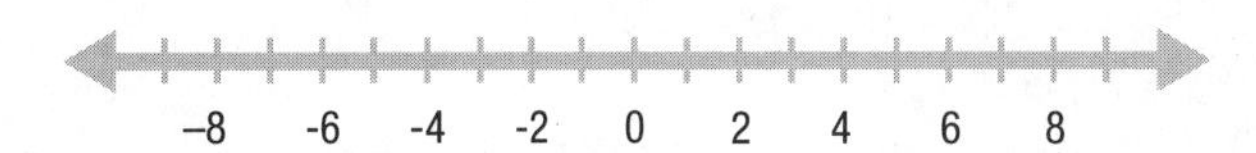


16. Factor: $6x + 3ax + 2b + ab$
17. Find the equation of the line through the point (–7, 12) with slope m = –2:
    a. in point-slope form.
    b. in slope-intercept form.
    c. in standard form.
18. Find:
    a. $-2x^0 - \frac{4}{y^0}$
    b. $\left(\frac{x^7yz^4}{x^2z}\right)^2$
    c. $a^0 \cdot a^0 \cdot a^0$
19. Solve for $x$ by factoring: $x^2 - 5x - 14 = 0$
20. Find:
    a. $11^3 \cdot 11^5$
    b. $\frac{x}{x^8}$
    c. $(ab^6)^3$

21. Find the equation of the line through the point (4, –3) with slope $-\frac{8}{5}$:

a. in point-slope form.

b. in slope-intercept form.

c. in standard form.

22. Factor: $4y^2 - 28y + 49$

23. Find the slope of the line through the points (31, 16) and (–2, 8).

24. Find: $(5y - 3)^2$

25. Solve $5y + 5 = 5(2 + y)$ for $y$.

26. Find the slope and $y$-intercept of the line $\frac{9}{4}x - \frac{2}{3}y = 2$.

27. Factor: $5x^2 + 2x - 7$

28. Find: $(xy^3 - 5x^2y + 11xy - 1) - (4xy^3 - 7 - x^2y + 3xy)$

29. Evaluate the expression $2a^3 - 8ab + 5b^2$ when $a = -2$ and $b = 4$.

30. Factor: $2x^2 + 9x - 18$

31. Use the FOIL method to find: $(a + 4)(a - 2)$

32. Graph the inequality $3x + 5y > 8$.

33. Jerome owed a total of $1820 on his two credit cards last year for which he paid $278.60 in interest. If one card charged 14% in interest and the other card charged 16% in interest, how much did he owe on each card?

34. Solve $-6 \leq 4y - 3 \leq 5$ for $y$.

35. Find the equation of the line through the point (5, 3) with slope $m = \frac{5}{6}$:

a. in point-slope form.

b. in slope-intercept form.

c. in standard form.

36. Factor: $x^2 - 7x + 12$

37. Graph the system of inequalities below to find its solution.

$$4x - 3y \leq 6$$

$$y \leq \frac{1}{2}x + 5$$

38. Circle the expressions below that are monomials.

| | |
|---|---|
| $15$ | $2x^4 + x$ |
| $9y$ | $a^3b^4c^2$ |

39. Graph the inequality $2x - 3y > 12$.

40. Hye was cleaning out her car and found a total of 44 nickels and quarters worth $4.80. How many of each did she find?

41. Factor: $36b^2 + 60b + 25$

42. Circle the true statements below.

$\frac{17}{21} - \frac{7}{18} = \frac{10}{3}$

$\frac{15}{33} = \frac{5}{11}$

$|8| - |13| = |8 - 13|$

The GCF of 120 and 252 is 12.

$11^2(15 - 2) = 121(15 - 2)$

43. Simplify this expression: $8x^3 + 5xy^2 + 7 - 4x^3 + 2xy^2$

44. Graph the system of inequalities below to find its solution.

$$x + 2y < 6$$

$$x + 2y \geq -5$$

45. Solve this system:

$$5x + 7y = 25$$

$$x - 3y = -17$$

46. Find the slope of the line through the points (2, 11) and (6, –8).

47. Find: $(x + 5)(5x + xy - 3y)$

48. Factor: $3x^2 - 5x - 2$

49. The length of a rectangle is 3 times its width. If the perimeter of the rectangle is 136 feet, what are its dimensions?

50. Solve this system:

$$6x + y = 1$$

$$2x + 5y = -9$$

# ANSWERS

## Cumulative Review Problems

**1.** $a^3 - 2a + 5$ **3.** $7x^2 + 5y + 2$

**5.** Alfredo should use 130 ml of the 15% solution and 120 ml of the 40% solution.

**7.** $(x-3)^2$ **9.** a, d, e

**11.**

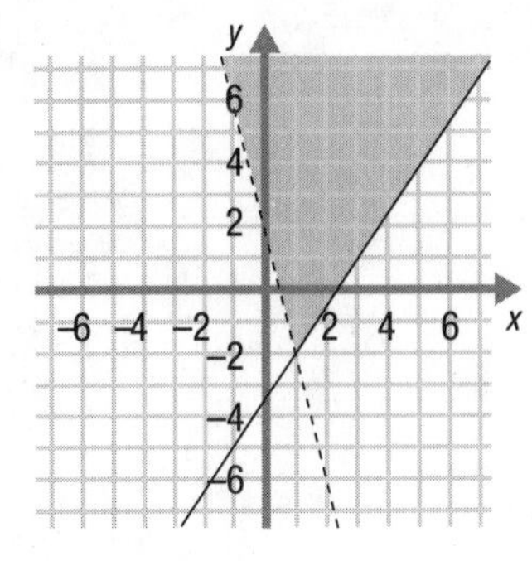


**13.** $(a^2 + 2)^2$

**15.** $6 \le x < 10$

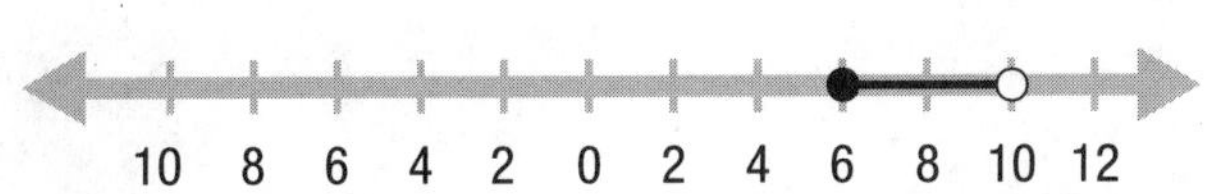


**17a.** $y - 12 = -2(x + 7)$

**b.** $y = -2x - 2$

**c.** $2x + y = -2$

**19.** $x = 7$ or $x = -2$

**21a.** $y + 3 = -\frac{8}{5}(x - 4)$

**b.** $y = -\frac{8}{5}x + \frac{17}{5}$

**c.** $8x + 5y = 17$

**23.** $\frac{8}{33}$ **25.** No solutions **27.** $(5x + 7)(x - 1)$ **29.** 128

**31.** $a^2 + 2a - 8$

**33.** Jerome owed $1,190 on the credit card that charged 16% interest and $630 on the credit card that charged 14% interest.

**35a.** $y - 3 = \frac{5}{6}(x - 5)$

**b.** $y = \frac{5}{6}x - \frac{7}{6}$

**c.** $-5x + 6y = -7$ or $5x - 6y = 7$

**37.**

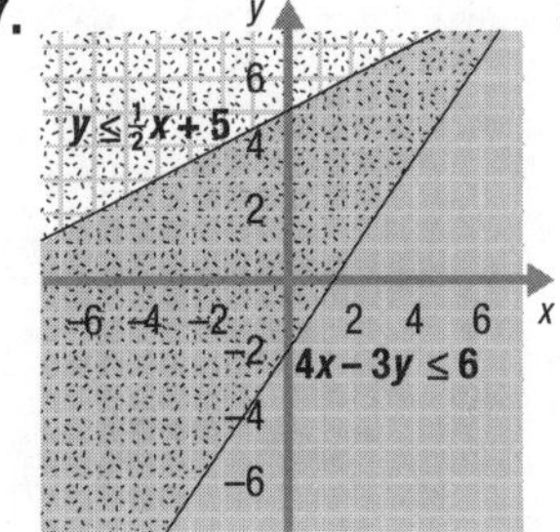


**39.**

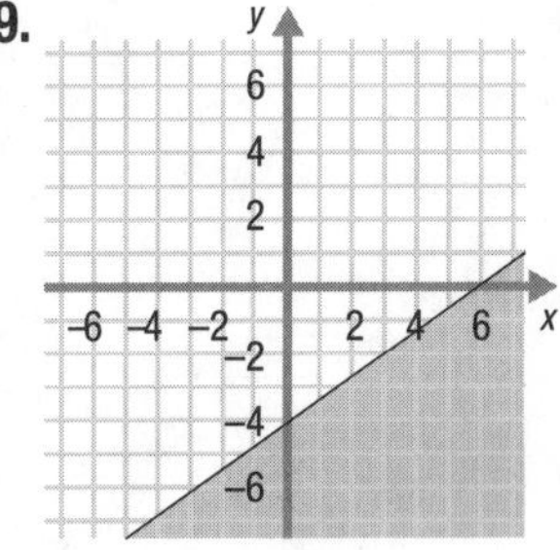


**41.** $(6b + 5)^2$ **43.** $4x^3 + 7xy^2 + 7$ **45.** $(-2, 5)$

**47.** $5x^2 + x^2y + 2xy + 25x - 15y$

**49.** The width of the rectangle is 17 feet and the length is 51 feet.

# TOPIC 7 INDEX

# LESSON 8.1 – RATIONAL EXPRESSIONS I

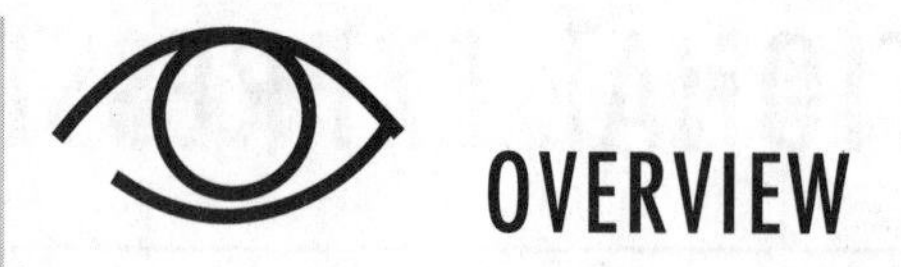

# OVERVIEW

***Here is what you'll learn in this lesson:***

**_Multiplying and Dividing_**

a. *Determining when a rational expression is undefined*

b. *Writing a rational expression in lowest terms*

c. *Multiplying rational expressions*

d. *Dividing rational expressions*

e. *Simplifying a complex fraction*

**_Adding and Subtracting_**

a. *Adding rational expressions with the same denominator*

b. *Subtracting rational expressions with the same denominator*

Almost 2000 years ago an Alexandrian mathematician named Eratosthenes figured out the circumference of the earth. He did this by setting up an equation that involved rational expressions, fractions in which the numerator and denominator are polynomials.

In this lesson, you will learn how to multiply, divide, add, and subtract rational expressions.

# EXPLAIN

## MULTIPLYING AND DIVIDING

### Summary

### Determining When a Rational Expression is Undefined

You have divided integers to form fractions, or rational numbers. Similarly, you can divide one polynomial by another to form an algebraic fraction. Algebraic fractions are also called rational expressions.

*A rational number can be written as a ratio of two integers, $\frac{a}{b}$, where $b \neq 0$.*

For example, here are some rational expressions:

$$\frac{5x^2 - 7x + 2}{x + 5} \qquad \frac{1}{7x^2 + 3} \qquad \frac{8x^2 - 9}{3x^5 + 4x - 1}$$

In general, a rational expression is written in the form $\frac{P}{Q}$ where $P$ and $Q$ are polynomials and $Q \neq 0$.

*A rational number such as $\frac{5}{19}$ is a special case of a rational expression. Here, the polynomials in the numerator and the denominator each consist of one term, an integer.*

A rational expression is undefined when the value of its denominator is zero. To find the values of the variable for which a rational expression is undefined:

1. Set the polynomial in the denominator equal to 0.
2. Solve the equation in step 1.

For example, to determine when $\frac{2x^2 + 5x - 6}{x - 4}$ is undefined:

1. Set the denominator equal to zero. $\quad x - 4 = 0$
2. Solve the equation $x - 4 = 0$ for $x$. $\quad x = 4$

So, $\frac{2x^2 + 5x - 6}{x - 4}$ is undefined when $x = 4$.

*It's okay if the numerator of a rational expression is zero. But the denominator can't be zero since division by 0 is undefined.*

### Writing a Rational Expression in Lowest Terms

You have learned how to reduce fractions to lowest terms. In a similar way, you can reduce rational expressions to lowest terms.

To reduce a rational expression to lowest terms:

1. Factor the numerator.
2. Factor the denominator.
3. Cancel pairs of factors that are common to both the numerator and the denominator.

*Remember, to reduce a fraction to lowest terms:*

1. *Find the prime factorization of the numerator.* $\quad \frac{6}{9} = \frac{2 \cdot 3}{9}$
2. *Find the prime factorization of the denominator.* $\quad = \frac{2 \cdot 3}{3 \cdot 3}$
3. *Cancel pairs of prime factors common to both the numerator and the denominator.* $\quad = \frac{2 \cdot \overset{1}{\cancel{3}}}{\underset{1}{\cancel{3}} \cdot 3}$

*So, $\frac{6}{9} = \frac{2}{3}$.*

For example, to reduce $\frac{21a^4b^3c^2}{3a^3cd^2}$ to lowest terms:

1. Factor the numerator. $= \frac{3 \cdot 7 \cdot a \cdot a \cdot a \cdot a \cdot b \cdot b \cdot b \cdot c \cdot c}{3a^3cd^2}$

2. Factor the denominator. $= \frac{3 \cdot 7 \cdot a \cdot a \cdot a \cdot a \cdot b \cdot b \cdot b \cdot c \cdot c}{3 \cdot a \cdot a \cdot a \cdot c \cdot d \cdot d}$

3. Cancel common factors. $= \frac{\overset{1}{\cancel{3}} \cdot 7 \cdot \overset{1}{\cancel{a}} \cdot \overset{1}{\cancel{a}} \cdot \overset{1}{\cancel{a}} \cdot a \cdot b \cdot b \cdot b \cdot \overset{1}{\cancel{c}} \cdot c}{\underset{1}{\cancel{3}} \cdot \underset{1}{\cancel{a}} \cdot \underset{1}{\cancel{a}} \cdot \underset{1}{\cancel{a}} \cdot \underset{1}{\cancel{c}} \cdot d \cdot d}$

$= \frac{7ab^3c}{d^2}$

As another example, to reduce $\frac{x^2 - 3x - 10}{x^2 - 2x - 15}$ to lowest terms:

1. Factor the numerator. $= \frac{(x + 2)(x - 5)}{x^2 - 2x - 15}$

2. Factor the denominator. $= \frac{(x + 2)(x - 5)}{(x + 3)(x - 5)}$

3. Cancel common factors. $= \frac{(x + 2)\overset{1}{\cancel{(x - 5)}}}{(x + 3)\underset{1}{\cancel{(x - 5)}}}$

$= \frac{x + 2}{x + 3}$

So, $\frac{x^2 - 3x - 10}{x^2 - 2x - 15} = \frac{x + 2}{x + 3}$.

*Remember, to multiply fractions:*

1. *Factor the numerators and denominators into prime factors.* $\frac{7}{18} \cdot \frac{9}{14} = \frac{7}{2 \cdot 3 \cdot 3} \cdot \frac{3 \cdot 3}{2 \cdot 7}$
2. *Cancel pairs of factors common to the numerators and denominators.* $= \frac{\overset{1}{\cancel{7}}}{2 \cdot \underset{1}{\cancel{3}} \cdot \underset{1}{\cancel{3}}} \cdot \frac{\overset{1}{\cancel{3}} \cdot \overset{1}{\cancel{3}}}{2 \cdot \underset{1}{\cancel{7}}}$
3. *Multiply the numerators and the denominators.* $= \frac{1}{4}$

## Multiplying Rational Expressions

You have learned how to multiply fractions. You can multiply rational expressions in the same way.

To multiply rational expressions:

1. Factor the numerators and denominators.
2. Cancel all pairs of factors common to the numerators and denominators.
3. Multiply the numerators. Multiply the denominators.

For example, to find $\frac{9x^2y^2}{4wz^2} \cdot \frac{8wz}{3x^2y}$:

1. Factor the numerators and denominators. $= \frac{3 \cdot 3 \cdot x \cdot x \cdot y \cdot y}{2 \cdot 2 \cdot w \cdot z \cdot z} \cdot \frac{2 \cdot 2 \cdot 2 \cdot w \cdot z}{3 \cdot x \cdot x \cdot y}$

2. Cancel pairs of factors common to the numerators and denominators. $= \frac{\overset{1}{\cancel{3}} \cdot 3 \cdot \overset{1}{\cancel{x}} \cdot \overset{1}{\cancel{x}} \cdot \overset{1}{\cancel{y}} \cdot y}{\underset{1}{\cancel{2}} \cdot \underset{1}{\cancel{2}} \cdot \underset{1}{\cancel{w}} \cdot \underset{1}{\cancel{z}} \cdot z} \cdot \frac{\overset{1}{\cancel{2}} \cdot \overset{1}{\cancel{2}} \cdot 2 \cdot \overset{1}{\cancel{w}} \cdot \overset{1}{\cancel{z}}}{\underset{1}{\cancel{3}} \cdot \underset{1}{\cancel{x}} \cdot \underset{1}{\cancel{x}} \cdot \underset{1}{\cancel{y}}}$

3. Multiply the numerators. Multiply the denominators. $= \frac{3 \cdot y}{z} \cdot \frac{2}{1}$

$= \frac{6y}{z}$

So, $\frac{9x^2y^2}{4wz^2} \cdot \frac{8wz}{3x^2y} = \frac{6y}{z}$.

## Dividing Rational Expressions

You have learned how to divide fractions. You can divide rational expressions in the same way.

To divide rational expressions:

1. Invert the second fraction and change "÷" to "·". Then multiply.
2. Factor the numerators and denominators.
3. Cancel pairs of factors common to the numerators and denominators.
4. Multiply the numerators. Multiply the denominators.

For example, to find $\frac{45xy^2}{2w^2} \div \frac{30x^2y^3}{w^3}$:

1. Invert the second fraction and change ÷ to ·. Then multiply. $= \frac{45xy^2}{2w^2} \cdot \frac{w^3}{30x^2y^3}$
2. Factor the numerators and denominators. $= \frac{3 \cdot 3 \cdot 5 \cdot x \cdot y \cdot y}{2 \cdot w \cdot w} \cdot \frac{w \cdot w \cdot w}{2 \cdot 3 \cdot 5 \cdot x \cdot x \cdot y \cdot y \cdot y}$
3. Cancel pairs of factors common to the numerator and denominator. $= \frac{\overset{1}{\cancel{3}} \cdot 3 \cdot \overset{1}{\cancel{5}} \cdot \overset{1}{\cancel{x}} \cdot \overset{1}{\cancel{y}} \cdot \overset{1}{\cancel{y}}}{2 \cdot \underset{1}{\cancel{w}} \cdot \underset{1}{\cancel{w}}} \cdot \frac{\overset{1}{\cancel{w}} \cdot \overset{1}{\cancel{w}} \cdot w}{2 \cdot \underset{1}{\cancel{3}} \cdot \underset{1}{\cancel{5}} \cdot \underset{1}{\cancel{x}} \cdot x \cdot \underset{1}{\cancel{y}} \cdot \underset{1}{\cancel{y}} \cdot y}$
4. Multiply the numerators. Multiply the denominators. $= \frac{3}{2} \cdot \frac{w}{2xy}$

$$= \frac{3w}{4xy}$$

So, $\frac{45xy^2}{2w^2} \div \frac{30x^2y^3}{w^3} = \frac{3w}{4xy}$.

*Remember, to divide fractions:*

1. *Invert the second fraction and change ÷ to ·. Then multiply.*

$$\frac{4}{27} \div \frac{8}{9} = \frac{4}{27} \cdot \frac{9}{8}$$

2. *Factor the numerators and denominators.*

$$= \frac{2 \cdot 2}{3 \cdot 3 \cdot 3} \cdot \frac{3 \cdot 3}{2 \cdot 2 \cdot 2}$$

3. *Cancel pairs of common factors.*

$$= \frac{\overset{1}{\cancel{2}} \cdot \overset{1}{\cancel{2}}}{\underset{1}{\cancel{3}} \cdot \underset{1}{\cancel{3}} \cdot 3} \cdot \frac{\overset{1}{\cancel{3}} \cdot \overset{1}{\cancel{3}}}{\underset{1}{\cancel{2}} \cdot \underset{1}{\cancel{2}} \cdot 2}$$

4. *Multiply the numerators. Multiply the denominators.*

$$= \frac{1}{3} \cdot \frac{1}{2}$$

$$= \frac{1}{6}$$

*So,* $\frac{4}{27} \div \frac{8}{9} = \frac{1}{6}$.

## Simplifying a Complex Fraction

When a fraction contains other fractions or rational expressions it is called a complex fraction. One way to simplify a complex fraction is to use division.

To simplify a complex fraction using division:

1. Rewrite the complex fraction as a division problem.
2. To divide, invert the second fraction and multiply.
3. Factor the numerators and denominators.
4. Cancel pairs of factors common to the numerator and denominator.
5. Multiply the numerators. Multiply the denominators.

For example, to simplify this complex fraction $\dfrac{\frac{3m^3}{5n^5}}{\frac{6m^2}{10n^2}}$:

1. Rewrite the complex fraction as a division problem. $= \dfrac{3m^3}{5n^5} \div \dfrac{6m^2}{10n^2}$

2. Divide by inverting the second fraction and multiplying. $= \dfrac{3m^3}{5n^5} \cdot \dfrac{10n^2}{6m^2}$

3. Factor the numerators and denominators. $= \dfrac{3 \cdot m \cdot m \cdot m}{5 \cdot n \cdot n \cdot n \cdot n \cdot n} \cdot \dfrac{2 \cdot 5 \cdot n \cdot n}{2 \cdot 3 \cdot m \cdot m}$

4. Cancel pairs of factors common to the numerator and denominator. $= \dfrac{\overset{1}{\cancel{3}} \cdot \overset{1}{\cancel{m}} \cdot \overset{1}{\cancel{m}} \cdot m}{\underset{1}{\cancel{5}} \cdot \underset{1}{\cancel{n}} \cdot \underset{1}{\cancel{n}} \cdot n \cdot n \cdot n} \cdot \dfrac{\overset{1}{\cancel{2}} \cdot \overset{1}{\cancel{5}} \cdot \overset{1}{\cancel{n}} \cdot \overset{1}{\cancel{n}}}{\underset{1}{\cancel{2}} \cdot \underset{1}{\cancel{3}} \cdot \underset{1}{\cancel{m}} \cdot \underset{1}{\cancel{m}}}$

5. Multiply the numerators. Multiply the denominators. $= \dfrac{m}{n^3}$

So, $\dfrac{\frac{3m^3}{5n^5}}{\frac{6m^2}{10n^2}} = \dfrac{m}{n^3}$.

## Sample Problems

***Answers to Sample Problems***

1. Find the values of $x$ for which this rational expression is undefined: $\dfrac{x^2-4}{(x+8)(x-5)}$

   ☑ a. Set the polynomial in the denominator equal to 0. $(x+8)(x-5)=0$

   ☐ b. Solve the equation $(x+8)(x-5)=0$. $x=$ ___ or $x=$ ___

*b. −8, 5 (in either order)*

2. Reduce to lowest terms: $\dfrac{24x^3yz^2}{6x^2z^3w}$

   ☑ a. Factor the numerator. $= \dfrac{2 \cdot 2 \cdot 2 \cdot 3 \cdot x \cdot x \cdot x \cdot y \cdot z \cdot z}{6x^2z^3w}$

   ☐ b. Factor the denominator. $= \dfrac{2 \cdot 2 \cdot 2 \cdot 3 \cdot x \cdot x \cdot x \cdot y \cdot z \cdot z}{\_ \cdot \_ \cdot \_ \cdot \_ \cdot \_ \cdot \_ \cdot \_ \cdot \_}$

   ☐ c. Cancel common factors. $=$ ______

*b. 2, 3, x, x, z, z, z, w*

*c. $\dfrac{4xy}{zw}$*

3. Reduce to lowest terms: $\dfrac{x^2+3x-4}{x^2-3x+2}$

   ☑ a. Factor the numerator. $= \dfrac{(x-1)(x+4)}{x^2-3x+2}$

   ☐ b. Factor the denominator. $= \dfrac{(x-1)(x+4)}{(\_\_\_)(\_\_\_)}$

   ☐ c. Cancel common factors. $= \dfrac{x+4}{\_\_\_}$

*b. x − 1, x − 2 (in either order)*

*c. x − 2*

4. Find: $\frac{3a^2b}{2cd^3} \cdot \frac{cd^2}{6a^3b^2}$

☑ a. Factor the numerators and denominators. $= \frac{3 \cdot a \cdot a \cdot b}{2 \cdot c \cdot d \cdot d \cdot d} \cdot \frac{c \cdot d \cdot d}{2 \cdot 3 \cdot a \cdot a \cdot a \cdot b \cdot b}$

☐ b. Cancel pairs of factors common to the numerators and denominators. $= \frac{1}{\_ \cdot \_} \cdot \frac{1}{\_ \cdot \_ \cdot \_}$

☐ c. Multiply the numerators. Multiply the denominators. $= \_\_\_\_$

5. Find: $\frac{6a^2b^2}{5c^3} \div \frac{3ab}{10c^2d}$

☑ a. Invert the second fraction and change ÷ to ·. Then multiply. $= \frac{6a^2b^2}{5c^3} \cdot \frac{10c^2d}{3ab}$

☐ b. Factor the numerators and denominators. $= \frac{2 \cdot 3 \cdot a \cdot a \cdot b \cdot b}{\_ \cdot \_ \cdot \_ \cdot \_} \cdot \frac{2 \cdot 5 \cdot c \cdot c \cdot d}{\_ \cdot \_ \cdot \_}$

☐ c. Cancel pairs of factors common to the numerator and denominator. $= \_\_\_ \cdot \_\_\_$

☐ d. Multiply the numerators. Multiply the denominators. $= \_\_\_$

6. Simplify this complex fraction: $\dfrac{\frac{a^3}{3b^4}}{\frac{7a^2}{6b^2}}$

☑ a. Rewrite as a division problem. $= \frac{a^3}{3b^4} \div \frac{7a^2}{6b^2}$

☐ b. Divide. (Invert the second fraction and multiply.) $= \frac{a^3}{3b^4} \cdot \_\_\_$

☐ c. Factor the numerators and denominators. $= \_\_\_\_\_\_$

☐ d. Cancel pairs of factors common to the numerator and denominator. $= \_\_\_\_\_\_$

☐ e. Multiply the numerators. Multiply the denominators. $= \_\_\_\_\_\_$

***Answers to Sample Problems***

*4. b. 2, d; 2, a, b*

*c.* $\frac{1}{4abd}$

*5. b. 5, c, c, c; 3, a, b*

*c.* $\frac{2ab}{c}, \frac{2d}{1}$ *or 2d*

*d.* $\frac{4abd}{c}$

*6. b.* $\frac{6b^2}{7a^2}$

*c.* $\frac{a \cdot a \cdot a}{3 \cdot b \cdot b \cdot b \cdot b} \cdot \frac{2 \cdot 3 \cdot b \cdot b}{7 \cdot a \cdot a}$

*d.* $\frac{a}{b \cdot b} \cdot \frac{2}{7}$

*e.* $\frac{2a}{7b^2}$

# ADDING AND SUBTRACTING

## Summary

### Adding Rational Expressions with the Same Denominator

*Remember, to add fractions with the same denominator:*

1. *Add the numerators. The denominator stays the same.*
$$\frac{3}{4} + \frac{7}{4} = \frac{3+7}{4} = \frac{10}{4}$$
2. *Factor the numerator and denominator.* $= \frac{\overset{1}{\cancel{2}} \cdot 5}{\underset{1}{\cancel{2}} \cdot 2}$
3. *Reduce to lowest terms.* $= \frac{5}{2}$

You have learned how to add fractions with the same denominator.
You can add rational expressions with the same denominator in a similar way.

To add rational expressions with the same denominator, add the numerators. The denominator stays the same.

For example, to find $\frac{3x}{x+5} + \frac{11}{x+5}$:

1. Add the numerators. $= \frac{3x+11}{x+5}$
   The denominator stays the same.

So, $\frac{3x}{x+5} + \frac{11}{x+5} = \frac{3x+11}{x+5}$.

After you add rational expressions, you often simplify the resulting rational expression by reducing it to lowest terms.

For example, to find $\frac{x+1}{x^2-3x-4} + \frac{2x+2}{x^2-3x-4}$:

1. Add the numerators. The denominator stays the same. $= \frac{x+1+2x+2}{x^2-3x-4}$
   $= \frac{3x+3}{x^2-3x-4}$
2. Factor the numerator and denominator. $= \frac{3\overset{1}{\cancel{(x+1)}}}{\underset{1}{\cancel{(x+1)}}(x-4)}$
3. Reduce to lowest terms. $= \frac{3}{x-4}$

So, $\frac{x+1}{x^2-3x-4} + \frac{2x+2}{x^2-3x-4} = \frac{3}{x-4}$.

### Subtracting Rational Expressions with the Same Denominator

*Remember, to subtract fractions with the same denominator:*

1. *Subtract the numerators. The denominator stays the same.*
$$\frac{5}{6} - \frac{1}{6} = \frac{5-1}{6} = \frac{4}{6}$$
2. *Factor the numerator and denominator.* $= \frac{\overset{1}{\cancel{2}} \cdot 2}{\underset{1}{\cancel{2}} \cdot 3}$
3. *Reduce to lowest terms.* $= \frac{2}{3}$

You have learned how to subtract fractions with the same denominator.
You can subtract rational expressions with the same denominator in a similar way.

To subtract rational expressions with the same denominator, subtract the numerators. The denominator stays the same.

For example, to find $\frac{4y}{17w} - \frac{5}{17w}$:

1. Subtract the numerators. $= \frac{4y-5}{17w}$
   The denominator stays the same.

So, $\frac{4y}{17w} - \frac{5}{17w} = \frac{4y-5}{17w}$.

After you subtract rational expressions, you often simplify the resulting rational expression by reducing it to lowest terms.

For example, to find $\frac{5x-6}{x^2+6x+5} - \frac{4x-7}{x^2+6x+5}$:

1. Subtract the numerators. The denominator stays the same. $= \frac{5x-6-(4x-7)}{x^2+6x+5}$

2. Distribute. Be careful with the signs. $= \frac{5x-6-4x+7}{x^2+6x+5}$

$= \frac{x+1}{x^2+6x+5}$

3. Factor the numerator and denominator. $= \frac{\cancel{x+1}^{\,1}}{\cancel{(x+1)}_{1}(x+5)}$

4. Reduce to lowest terms. $= \frac{1}{x+5}$

So, $\frac{5x-6}{x^2+6x+5} - \frac{4x-7}{x^2+6x+5} = \frac{1}{x+5}$.

## Sample Problems

1. Find: $\frac{2a}{5b} + \frac{6a}{5b}$

 ☐ a. Add the numerators. The denominator stays the same. $= \frac{\rule{2em}{0.4pt}}{5b}$

 $= \rule{3em}{0.4pt}$

2. Find: $\frac{2x}{x-1} + \frac{13}{x-1}$:

 ☐ a. Add the numerators. The denominator stays the same. $= \frac{\rule{2em}{0.4pt}}{x-1}$

3. Find: $\frac{x-12}{x^2-2x-15} + \frac{3x-8}{x^2-2x-15}$

 ☐ a. Add the numerators. The denominator stays the same. $= \frac{(x-12)+(3x-8)}{x^2-2x-15}$

 $= \frac{\rule{3em}{0.4pt}}{x^2-2x-15}$

 ☐ b. Factor the numerator and denominator. $= \rule{4em}{0.4pt}$

 ☐ c. Reduce to lowest terms. $= \rule{4em}{0.4pt}$

***Answers to Sample Problems***

*a.* $2a + 6a$

$\frac{8a}{5b}$

*a.* $2x + 13$

*a.* $4x - 20$

*b.* $\frac{4(x-5)}{(x+3)(x-5)}$

*c.* $\frac{4}{x+3}$

**Answers to Sample Problems**

a. $7 - 2$

$\frac{5}{11y}$

a. $13 - (2 + x)$

b. $13 - 2 - x$

$\frac{11 - x}{4x}$

a. $\frac{x - 5}{x^2 - 25}$

b. $\frac{x - 5}{(x + 5)(x - 5)}$

c. $\frac{1}{x + 5}$

4. Find: $\frac{7}{11y} - \frac{2}{11y}$

☐ a. Subtract the numerators. The denominator stays the same. $= \frac{\quad}{11y}$

$=$ ____

5. Find: $\frac{13}{4x} - \frac{2 + x}{4x}$

☐ a. Subtract the numerators. The denominator stays the same. $= \frac{\quad}{4x}$

☐ b. Distribute. Be careful with the signs. $= \frac{\quad}{4x}$

$=$ ____

6. Find: $\frac{5x + 2}{x^2 - 25} - \frac{4x + 7}{x^2 - 25}$

☑ a. Subtract the numerators. The denominator stays the same. $= \frac{5x + 2 - (4x + 7)}{x^2 - 25}$

$=$ ____

☐ b. Factor the numerator and denominator. $=$ ____

☐ c. Reduce to lowest terms. $=$ ____

# HOMEWORK

## Homework Problems

Circle the homework problems assigned to you by the computer, then complete them below.

### Explain

### Multiplying and Dividing

1. For what values of $x$ is the rational expression below undefined?

   $$\frac{(x+7)(x-8)}{(x-14)(x+2)}$$

2. For what values of $x$ is the rational expression below undefined?

   $$\frac{x^2-9}{x^2-4}$$

3. Reduce to lowest terms: $\frac{3a^2b^5}{27ab^7}$
4. Reduce to lowest terms: $\frac{x^2-3x-28}{x^2+5x+4}$
5. Find: $\frac{3y^3}{z} \cdot \frac{yz}{7y^2}$
6. Find: $\frac{15a^2b^2}{2c^3d} \cdot \frac{2cd^3}{5ab^2}$
7. Find: $\frac{12xy}{w^2} \div \frac{4xy^3}{w^3}$
8. Find: $\frac{3ab^2}{13d^4} \div \frac{6a^2b}{11d^2}$
9. Simplify this complex fraction: $\frac{\frac{5}{a^3}}{\frac{9}{a^2}}$. Write your answer in lowest terms.
10. The ratio of the area of a circle to its circumference is given by $\frac{\pi r^2}{2\pi r}$.
    a. What value of $r$ makes this ratio undefined?
    b. Simplify this expression and then determine what value of $r$ will make the ratio equal to 2. That is, find the radius that will yield an area that is twice the circumference.
11. Simplify this complex fraction: $\frac{\frac{x}{x-1}}{\frac{1}{x+1}}$. Write your answer in lowest terms.
12. Simplify this complex fraction: $\frac{\frac{3y^4}{y+2}}{\frac{9y^2}{y-2}}$. Write your answer in lowest terms.

### Adding and Subtracting

13. Find: $\frac{3x}{5y} + \frac{18x}{5y}$
14. Find: $\frac{3+5a}{8-a} + \frac{2a+1}{8-a}$
15. Find: $\frac{x}{x^2-4} + \frac{4}{x^2-4}$
16. Find: $\frac{3z+4}{z-11} + \frac{2z}{z-11}$
17. Find: $\frac{2x-5}{x^2-5x-14} + \frac{3x+15}{x^2-5x-14}$
18. Find: $\frac{2z+11}{z^2-3z-18} + \frac{3z+4}{z^2-3z-18}$
19. Find: $\frac{9}{15x} - \frac{2}{15x}$
20. Find: $\frac{17}{13y} - \frac{5+y}{13y}$
21. The volume, $V$, of a sphere of radius $r$ is defined by the formula $V = \frac{4\pi r^3}{3}$. Find the volume of two identical spheres. That is, find $\frac{4\pi r^3}{3} + \frac{4\pi r^3}{3}$.
22. Find: $\frac{y}{y^2-81} - \frac{9}{y^2-81}$
23. Find: $\frac{4y+6}{3y+6} - \frac{3y+4}{3y+6}$
24. Find: $\frac{4x-4}{x^2-2x-15} - \frac{3x-7}{x^2-2x-15}$

# APPLY

## Practice Problems

Here are some additional practice problems for you to try.

### Multiplying and Dividing

1. For what value(s) of $x$ is the rational expression below undefined?
$\frac{1}{x+5}$

2. For what value(s) of $x$ is the rational expression below undefined?
$\frac{2}{(x-3)(x+5)}$

3. For what value(s) of $x$ is the rational expression below undefined?
$\frac{25}{3x^2-12}$

4. For what value(s) of $x$ is the rational expression below undefined?
$\frac{17}{2x^2-18}$

5. Reduce to lowest terms: $\frac{36m^5n^3}{27mn^6}$

6. Reduce to lowest terms: $\frac{75xy^2z^7}{45x^4y^3z^6}$

7. Reduce to lowest terms: $\frac{44a^2b^4c}{77a^7bc}$

8. Reduce to lowest terms: $\frac{x^2+10x+21}{x^2+5x+6}$

9. Reduce to lowest terms: $\frac{x^2+4x-5}{x^2-3x+2}$

10. Reduce to lowest terms: $\frac{x^2-x-12}{x^2+5x+6}$

11. Find: $\frac{6a}{b^3c^2} \cdot \frac{7b}{3a^3}$

12. Find: $\frac{8x}{y^2z} \cdot \frac{5y}{4x^2}$

13. Find: $\frac{10m^3n^5}{9mn^3} \cdot \frac{21m^5}{15n^6}$

14. Find: $\frac{5ab^4}{3c^2} \cdot \frac{6c}{10b^3}$

15. Find: $\frac{8m^5n}{7p^2} \cdot \frac{14mp}{24m^2n^4}$

16. Find: $\frac{3xy^3}{4z} \cdot \frac{2z^2}{9xy^2}$

17. Find: $\frac{4a^2b}{5c^3} \div \frac{8ab^2}{15c}$

18. Find: $\frac{12m^3n^4}{7p} \div \frac{18mn^5}{21p^4}$

19. Find: $\frac{3xy^2}{7z} \div \frac{6x^2y}{14z^2}$

20. Find: $\frac{5x^2y}{12z^3w} \div \frac{xy}{4z}$

21. Find: $\frac{9m^3n^4}{11pq^2} \div \frac{12mn^3}{22p^2q}$

22. Find: $\frac{7a^2b}{9c^2d^2} \div \frac{ab}{3cd}$

23. Simplify the complex fraction below.
$\frac{\frac{4m^2}{n^3}}{\frac{2m}{n}}$

24. Simplify the complex fraction below.
$\frac{\frac{6x^5}{5y^3}}{\frac{3x^3}{10y}}$

25. Simplify the complex fraction below.
$\frac{\frac{6a^2}{b^3}}{\frac{3a}{2b}}$

26. Simplify the complex fraction below.
$\frac{\frac{6a^3}{a+5}}{\frac{12a}{a-4}}$

27. Simplify the complex fraction below.
$\frac{\frac{5x^2}{x+7}}{\frac{10x}{x-3}}$

28. Simplify the complex fraction below.
$\frac{\frac{15y^5}{y-3}}{\frac{18y^3}{y+3}}$

## Adding and Subtracting

29. Find: $\frac{3a}{7b} + \frac{2a}{7b}$

30. Find: $\frac{2x}{5y} + \frac{7x}{5y}$

31. Find: $\frac{3b}{2b+1} + \frac{5}{2b+1}$

32. Find: $\frac{9n}{4n-7} + \frac{2}{4n-7}$

33. Find: $\frac{7x}{5x-1} + \frac{2}{5x-1}$

34. Find: $\frac{5y+2}{3y-2} - \frac{4y-5}{3y-2}$

35. Find: $\frac{7b+1}{2b+9} - \frac{5b+5}{2b+9}$

36. Find: $\frac{6x+1}{2x+3} - \frac{5x-3}{2x+3}$

37. Find: $\frac{15}{7n} - \frac{4-5n}{7n}$

38. Find: $\frac{11}{9x} - \frac{7-5x}{9x}$

39. Add and reduce your answer to lowest terms:
$\frac{3x+4}{x^2+5x+6} + \frac{5}{x^2+5x+6}$

40. Add and reduce your answer to lowest terms:
$\frac{4x+7}{x^2+7x+10} + \frac{1}{x^2+7x+10}$

41. Add and reduce your answer to lowest terms:
$\frac{2x+5}{x^2+x-12} + \frac{3}{x^2+x-12}$

42. Add and reduce your answer to lowest terms:
$\frac{3x+12}{x^2+2x-3} + \frac{2x+3}{x^2+2x-3}$

43. Add and reduce your answer to lowest terms:
$\frac{3x+19}{x^2+3x-10} + \frac{x+1}{x^2+3x-10}$

44. Add and reduce your answer to lowest terms:
$\frac{2x+15}{x^2+5x-14} + \frac{x+6}{x^2+5x-14}$

45. Add and reduce your answer to lowest terms:
$\frac{x^2-5x+2}{x^2+7x+12} + \frac{2(5x+1)}{x^2+7x+12}$

46. Add and reduce your answer to lowest terms:
$\frac{x^2-3x+2}{x^2+3x+2} + \frac{4(2x+1)}{x^2+3x+2}$

47. Add and reduce your answer to lowest terms:
$\frac{x^2-7x+12}{x^2+9x+18} + \frac{5(3x+1)-2}{x^2+9x+18}$

48. Subtract and reduce your answer to lowest terms:
$\frac{4x+7}{3x-9} - \frac{3x+10}{3x-9}$

49. Subtract and reduce your answer to lowest terms:
$\frac{5x+6}{2x+10} - \frac{2x-9}{2x+10}$

50. Subtract and reduce your answer to lowest terms:
$\frac{3x+2}{2x+14} - \frac{2x-5}{2x+14}$

51. Subtract and reduce your answer to lowest terms:
$\frac{2x+5}{x^2+3x-10} - \frac{x+7}{x^2+3x-10}$

52. Subtract and reduce your answer to lowest terms:
$\frac{4x+2}{x^2-x-20} - \frac{3x-2}{x^2-x-20}$

53. Subtract and reduce your answer to lowest terms:
$\frac{3x+5}{x^2-4x+3} - \frac{2x+8}{x^2-4x+3}$

54. Subtract and reduce your answer to lowest terms:
$\frac{8x+5}{4x-4} - \frac{5x+8}{4x-4}$

55. Subtract and reduce your answer to lowest terms:
$\frac{5x+9}{5x+20} - \frac{2x-3}{5x+20}$

56. Subtract and reduce your answer to lowest terms:
$\frac{7x-2}{3x+3} - \frac{5x-4}{3x+3}$

# EVALUATE

## Practice Test

Take this practice test to be sure that you are prepared for the final quiz in Evaluate.

1. For what values of $x$ is the following expression undefined? $\frac{x^2 - 16}{(x+3)(x-2)}$

2. Reduce to lowest terms: $\frac{x^2 - 3x - 28}{x^2 + 10x + 24}$

3. Find:

   a. $\frac{5y^2}{9z^2} \cdot \frac{z}{y^2}$

   b. $\frac{12x^2y^2}{z^3w} \cdot \frac{2zw}{3xy^2}$

4. a. Find: $\frac{3x^2}{yz} \div \frac{7x}{2yz}$

   b. Simplify this complex fraction: $\frac{\frac{2x^2y}{9w}}{\frac{10xy^2}{3w^2}}$. Write your answer in lowest terms.

5. Find:

   a. $\frac{5x}{13y} + \frac{2x}{13y}$

   b. $\frac{3w}{z-8} + \frac{14}{z-8}$

6. Find: $\frac{15y}{7x} - \frac{3+6y}{7x}$

7. Find the following. Reduce your answer to lowest terms.

   $$\frac{x+7}{x^2 - 3x - 18} + \frac{3x+5}{x^2 - 3x - 18}$$

8. Find the following. Reduce your answer to lowest terms.

   $$\frac{9y}{y-7} - \frac{3y-4}{y-7}$$

# ANSWERS

## Homework

**1.** $x = 14$ or $x = -2$ **3.** $\frac{a}{9b^2}$ **5.** $\frac{3y^2}{7}$ **7.** $\frac{3w}{y^2}$ **9.** $\frac{5}{9a}$

**11.** $\frac{x(x+1)}{x-1}$ or $\frac{x^2+x}{x-1}$ **13.** $\frac{21x}{5y}$ **15.** $\frac{x+4}{x^2-4}$ or $\frac{x+4}{(x+2)(x-2)}$

**17.** $\frac{5}{x-7}$ **19.** $\frac{7}{15x}$ **21.** $\frac{8\pi r^3}{3}$ **23.** $\frac{1}{3}$

## Practice Problems

**1.** $-5$ **3.** $2, -2$ **5.** $\frac{4m^4}{3n^3}$ **7.** $\frac{4b^3}{7a^5}$ **9.** $\frac{x+5}{x-2}$ **11.** $\frac{14}{a^2b^2c^2}$

**13.** $\frac{14m^7}{9n^4}$ **15.** $\frac{2m^4}{3n^3p}$ **17.** $\frac{3a}{2bc^2}$ **19.** $\frac{yz}{x}$ **21.** $\frac{3m^2np}{2q}$ **23.** $\frac{2m}{n^2}$

**25.** $\frac{4a}{b^2}$ **27.** $\frac{x^2-3x}{2x+14}$ or $\frac{x(x-3)}{2(x+7)}$ **29.** $\frac{5a}{7b}$ **31.** $\frac{3b+5}{2b+1}$

**33.** $\frac{7x+2}{5x-1}$ **35.** $\frac{2b-4}{2b+9}$ **37.** $\frac{11+5n}{7n}$ **39.** $\frac{3}{x+2}$ **41.** $\frac{2}{x-3}$

**43.** $\frac{4}{x-2}$ **45.** $\frac{x+1}{x+3}$ **47.** $\frac{x+5}{x+6}$ **49.** $\frac{3}{2}$

**51.** $\frac{1}{x+5}$ **53.** $\frac{1}{x-1}$ **55.** $\frac{3}{5}$

## Practice Test

**1.** $x = -3$ or $x = 2$ **2.** $\frac{x-7}{x+6}$

**3a.** $\frac{5}{9z}$ **b.** $\frac{8x}{z^2}$

**4a.** $\frac{6x}{7}$ **b.** $\frac{xw}{15y}$

**5a.** $\frac{7x}{13y}$ **b.** $\frac{3w+14}{z-8}$

**6.** $\frac{9y-3}{7x}$ **7.** $\frac{4}{x-6}$ **8.** $\frac{6y+4}{y-7}$

# LESSON 8.2 – RATIONAL EXPRESSIONS II

# OVERVIEW

*Here is what you'll learn in this lesson:*

***Negative Exponents***

a. *Notation*

b. *Scientific notation*

***Multiplying and Dividing***

a. *Reducing a rational expression of the form* $\frac{a-b}{b-a}$

b. *Multiplying rational expressions*

c. *Dividing rational expressions*

d. *Simplifying a complex fraction*

***Adding and Subtracting***

a. *Finding the least common denominator of rational expressions*

b. *Adding rational expressions with different denominators*

c. *Subtracting rational expressions with different denominators*

d. *Simplifying a complex fraction*

Rational expressions are fractions in which the numerator and denominator are polynomials.

In this lesson, you will learn more about how to multiply, divide, add, and subtract rational expressions. You will also learn about negative exponents.

# EXPLAIN

## NEGATIVE EXPONENTS

### Summary

#### Negative Exponents

You have seen how to work with exponents that are positive integers or 0. Now you will learn about negative integer exponents.

For example, you have previously found $\frac{3^5}{3^7}$ by canceling:

$$\frac{3^5}{3^7} = \frac{\overset{1}{\cancel{3}} \cdot \overset{1}{\cancel{3}} \cdot \overset{1}{\cancel{3}} \cdot \overset{1}{\cancel{3}} \cdot \overset{1}{\cancel{3}}}{\underset{1}{\cancel{3}} \cdot \underset{1}{\cancel{3}} \cdot \underset{1}{\cancel{3}} \cdot \underset{1}{\cancel{3}} \cdot \underset{1}{\cancel{3}} \cdot 3 \cdot 3} = \frac{1}{3^2}$$

You can also find $\frac{3^5}{3^7}$ by subtracting exponents:

$$\frac{3^5}{3^7} = 3^{5-7} = 3^{-2}$$

So, $3^{-2} = \frac{1}{3^2}$.

In general:

$$x^{-n} = \frac{1}{x^n}$$

Here, $x \neq 0$ and $n$ is a nonnegative integer.

You can rewrite an expression with a negative exponent as a fraction by following these steps:

1. Write the numerator as 1.
2. Write the denominator as the original expression, except change the negative exponent to a positive exponent.
3. Do the multiplication in the denominator.

For example, to find $5^{-2}$:

1. Write the numerator as 1.
2. Write the denominator as the original expression, except change the negative exponent to a positive exponent. $\frac{1}{5^2}$
3. Do the multiplication in the denominator. $= \frac{1}{25}$

*Even though the 5 is raised to a negative power, the result is a positive number.*

So, $5^{-2} = \frac{1}{25}$.

Similarly, here's how to simplify an expression raised to a negative exponent that is in the denominator of a fraction:

1. Take the expression out of the denominator and change the negative exponent to a positive exponent.

2. Do the multiplication.

For example, to find $\frac{1}{4^{-3}}$:

1. Take the expression out of the denominator and change the negative exponent to a positive exponent. $4^3$

2. Do the multiplication. $= 64$

*Why does this work?*

*Well, since $4^{-3} = \frac{1}{4^3}$, then*

$$\frac{1}{4^{-3}} = \frac{1}{\frac{1}{4^3}}$$
$$= 1 \div \frac{1}{4^3}$$
$$= 1 \cdot 4^3$$
$$= 64$$

In general:

$$\frac{1}{x^{-n}} = x^n$$

Here, $x \neq 0$ and $n$ is a nonnegative integer.

## Properties of Negative Exponents

The basic properties of nonnegative exponents also hold for negative exponents. The table below summarizes some of these properties.

| Property | Positive Integer Exponents | Negative Integer Exponents |
|---|---|---|
| Multiplication | $3^2 \cdot 3^4 = 3^{2+4} = 3^6$ | $3^{-2} \cdot 3^{-4} = 3^{(-2)+(-4)} = 3^{-6}$ |
| Division | $\frac{4^7}{4^5} = 4^{7-5} = 4^2$ | $\frac{4^{-7}}{4^{-5}} = 4^{(-7)-(-5)} = 4^{-7+5} = 4^{-2}$ |
| Power of a Power | $(5^2)^3 = 5^{2 \cdot 3} = 5^6$ | $(5^{-2})^{-3} = 5^{(-2)(-3)} = 5^6$ |
| Power of a Product | $(5 \cdot 7)^3 = 5^3 \cdot 7^3$ | $(5 \cdot 7)^{-3} = 5^{-3} \cdot 7^{-3}$ |
| Power of a Quotient | $\left(\frac{3}{5}\right)^4 = \frac{3^4}{5^4}$ | $\left(\frac{3}{5}\right)^{-4} = \frac{3^{-4}}{5^{-4}}$ |

Some other properties that hold for negative exponents are given below.

One property involving negative exponents shows how to rewrite a fraction in which the numerator and denominator are each raised to a negative power.

For example, here's one way to rewrite $\frac{2^{-4}}{5^{-3}}$ using only positive exponents:

$$\frac{2^{-4}}{5^{-3}} = \frac{\frac{1}{2^4}}{\frac{1}{5^3}}$$
$$= \frac{5^3}{2^4}$$
$$= \frac{125}{16}$$

*Notice that the $2^{-4}$ in the numerator became the $2^4$ in the denominator and that the $5^{-3}$ in the denominator became the $5^3$ in the numerator.*

In general:

$$\frac{x^{-m}}{y^{-n}} = \frac{y^n}{x^m}$$

Here, $x \neq 0$, $y \neq 0$, and $m$ and $n$ are nonnegative integers.

Another property involving negative exponents shows how to rewrite a fraction that is raised to a negative power.

For example, here's one way to simplify $\left(\frac{2}{3}\right)^{-4}$:

$$\left(\frac{2}{3}\right)^{-4} = \frac{2^{-4}}{3^{-4}}$$

$$= \frac{3^4}{2^4}$$

$$= \left(\frac{3}{2}\right)^4$$

*When a fraction is raised to a negative power, flip the fraction and change the negative power to positive.*

In general:

$$\left(\frac{x}{y}\right)^{-n} = \left(\frac{y}{x}\right)^n$$

Here, $x \neq 0$, $y \neq 0$, and $n$ is a nonnegative integer.

## Simplifying Expressions with Negative Exponents

You can combine different properties to simplify more complicated expressions containing negative exponents.

For example, to rewrite $(3r^4s^{-5}t)^{-2}$ using only positive exponents:

1. Use the Power of a Product Property. $(3r^4s^{-5}t)^{-2} = (3)^{-2} \cdot (r^4)^{-2} \cdot (s^{-5})^{-2} \cdot (t)^{-2}$

2. Use the Power of a Power Property. $= 3^{-2} \cdot r^{-8} \cdot s^{10} \cdot t^{-2}$
3. Rewrite the negative exponents as positive exponents. $= \frac{1}{3^2} \cdot \frac{1}{r^8} \cdot s^{10} \cdot \frac{1}{t^2}$
4. Simplify. $= \frac{s^{10}}{9r^8t^2}$

So, $(3r^4s^{-5}t)^{-2} = \frac{s^{10}}{9r^8t^2}$.

As another example, rewrite $\frac{m^3n^{-6}}{6m^{-2}n^3}$ using only positive exponents:

1. Since $\frac{x^{-m}}{y^{-n}} = \frac{y^n}{x^m}$, rewrite $\frac{n^{-6}}{m^{-2}}$ as $\frac{m^2}{n^6}$. $\frac{m^3n^{-6}}{6m^{-2}n^3} = \frac{m^3 \cdot m^2}{6n^6 \cdot n^3}$
2. Use the Multiplication Property of Exponents. $= \frac{m^5}{6n^9}$

So, $\frac{m^3n^{-6}}{6m^{-2}n^3} = \frac{m^5}{6n^9}$.

As a third example, to simplify $\frac{73}{3^{-2} + 4^{-3}}$:

1. Rewrite using positive exponents. $\frac{73}{3^{-2} + 4^{-3}} = \frac{73}{\frac{1}{3^2} + \frac{1}{4^3}}$

2. Multiply to eliminate the exponents. $= \frac{73}{\frac{1}{9} + \frac{1}{64}}$

3. Rewrite each fraction in the denominator with the LCD, 576. $= \frac{73}{\frac{64}{576} + \frac{9}{576}}$

4. Add the fractions in the denominator. $= \frac{73}{\frac{73}{576}}$

5. Simplify.

$$= 73 \div \frac{73}{576}$$

$$= 73 \cdot \frac{576}{73}$$

$$= 576$$

So, $\frac{73}{3^{-2} + 4^{-3}} = 576.$

## Scientific Notation

Now that you know how to work with negative exponents as well as positive exponents, you can learn how to use scientific notation. Scientific notation is a shorthand that is often used to write very small or very large numbers.

For example here are some very small and very large numbers:

- the gross national debt at the end of 1994 was $4,643,700,000,000
- a number called Planck's constant that relates the energy of a photon to its frequency is 0.000000000000004135 electron volt seconds
- the speed of light is 299,792,500 meters per second

You can rewrite these numbers in scientific notation.

A number written in scientific notation is the product of a number between 1 and 10 and an integer power of 10. To write a number in scientific notation:

*Remember the decimal point is at the end of a whole number. For example, the number 239 can be written like this: 239.0*

1. Move the decimal point to the left or to the right until you have a number between 1 and 10.
2. Multiply this number by a power of 10. To find the power, count the number of places you moved the decimal point.

   - If you moved the decimal point to the left, the power is positive.
   - If you moved the decimal point to the right, the power is negative.

For example, to write \$4,643,700,000,000, in scientific notation:

1. Move the decimal point to the left. — 4.643700000000 (12 11 10 9 8 7 6 5 4 3 2 1)

2. Multiply by the appropriate power of 10. — $4.6437 \times 10^{12}$

So, \$4,643,700,000,000 = $4.6437 \times 10^{12}$ dollars.

*You moved the decimal point 12 places to the left, so multiply by $10^{12}$.*

As another example. To write 0.000000000000004135 eVsec (electron volt seconds), in scientific notation:

1. Move the decimal point to the right. — 0.000000000000004.135 (1 2 3 4 5 6 7 8 9 10 11 12 13 14 15)

2. Multiply by the appropriate power of 10. — $4.135 \times 10^{-15}$

So, 0.000000000000004135 eVsec = $4.135 \times 10^{-15}$ eVsec.

*Here's a hint to help you know if you've moved the decimal point in the correct direction. If the power is positive, the number should be big. If the power is negative, the number should be less than 1.*

You can also reverse this process to write a number given in scientific notation in expanded form.

To write a number in expanded form you must perform the multiplication indicated by the power of 10 in scientific notation. So:

1. Move the decimal point the number of places indicated by the power of 10.
   - If the power is positive, move the decimal point to the right.
   - If the power is negative, move the decimal point to the left.

2. Fill in additional zeros as needed.

For example, to write $2.997925 \times 10^8 \frac{m}{s}$, in expanded form:

1. Move the decimal point to the right 8 places. — 2997925 . (1 2 3 4 5 6 7 8)

2. Fill in additional zeros. — 299792500. (1 2 3 4 5 6 7 8)

So, $2.997925 \times 10^8 \frac{m}{s} = 299{,}792{,}500 \frac{m}{s}$.

## Answers to Sample Problems

c. $\frac{1}{a^6}, \frac{1}{c^3}$

d. $\frac{125b^{21}}{a^6c^3}$

a. $y^3, x^4$

b. $\frac{3y^6}{x^6}$

b. $\frac{17}{\frac{1}{8}+\frac{1}{9}}$

c. $\frac{17}{\frac{9}{72}+\frac{8}{72}}$

d. $\frac{17}{\frac{17}{72}}$

e. $17 \div \frac{17}{72}$

$17 \cdot \frac{72}{17}$

$72$

b. $4.5318 \times 10^7$

a. 968

b. 0.0000968

## Sample Problems

1. Rewrite $(5a^{-2}b^7c^{-1})^3$ using only positive exponents.

- ☑ a. Use the Power of a Product Property. $= (5)^3 \cdot (a^{-2})^3 \cdot (b^7)^3 \cdot (c^{-1})^3$
- ☑ b. Use the Power of a Power Property. $= 125 \cdot a^{-6} \cdot b^{21} \cdot c^{-3}$
- ☐ c. Rewrite using only positive exponents. $= 125 \cdot$ ___ $\cdot\, b^{21} \cdot$ ___
- ☐ d. Simplify. = ______

2. Rewrite $\frac{3x^{-4}y^3}{x^2y^{-3}}$ using only positive exponents.

- ☐ a. Use $\frac{x^{-m}}{y^{-n}} = \frac{y^n}{x^m}$ to rewrite $\frac{x^{-4}}{y^{-3}}$. $\frac{3x^{-4}y^3}{x^2y^{-3}} = \frac{3\_\_\_ \cdot y^3}{x^2 \cdot \_\_\_}$
- ☐ b. Use the Multiplication Property of Exponents. = ______

3. Simplify this expression: $\frac{17}{2^{-3} + 3^{-2}}$

- ☑ a. Rewrite using only positive exponents. $= \frac{17}{\frac{1}{2^3}+\frac{1}{3^2}}$
- ☐ b. Multiply to eliminate the exponents. = ______
- ☐ c. Rewrite each fraction in the denominator with the LCD. = ______
- ☐ d. Add the fractions in the denominator. = ______
- ☐ e. Simplify. = ______

= ______

= ______

4. Write 45,318,000 in scientific notation.

- ☑ a. Move the decimal point. 4.5318000
- ☐ b. Multiply by the appropriate power of 10. = ______

5. Write $9.68 \times 10^{-5}$ in expanded form.

- ☐ a. Move the decimal point. ______
- ☐ b. Fill in additional zeros. = ______

# MULTIPLYING AND DIVIDING

## Summary

### Reducing a Rational Expression of the Form $\frac{a-b}{b-a}$

Look at this rational expression: $\frac{7-x}{x-7}$

Here, the numerator and denominator are already factored, and at first glance the rational expression may appear to be in lowest terms.

But notice that the numerator and denominator are the same except for their signs. To reduce this rational expresson further:

1. Factor –1 out of **either** the numerator or the denominator. Here, the numerator is factored. $\frac{7-x}{x-7} = \frac{-1(-7+x)}{x-7}$
2. Rewrite $(-7+x)$ as $(x-7)$. $= \frac{-1(x-7)}{x-7}$
3. Cancel common factors. $= \frac{-1\cancel{(x-7)}^{1}}{\cancel{x-7}_{1}}$

$= -1$

So, $\frac{7-x}{x-7} = -1$.

In general:

$$\frac{a-b}{b-a} = -1$$

Here, $a \neq b$.

Here's another example. To reduce $\frac{x^2+6x-27}{9-x^2}$ to lowest terms:

1. Factor the numerator and denominator. $= \frac{(x-3)(x+9)}{(3-x)(3+x)}$

   In the numerator, notice $(x-3)$.
   In the denominator, notice $(3-x)$.

2. Recall that $\frac{x-3}{3-x} = -1$. $= -1 \cdot \frac{(x+9)}{(3+x)}$

   So, you can rewrite the expression. $= -\frac{x+9}{3+x}$

## Multiplying Rational Expressions

You have learned how to multiply rational expressions. Here is an example.

Find $\dfrac{x^2 + 2x - 3}{x^2 + 5x + 6} \cdot \dfrac{x^2 + x - 12}{x^2 - 4x + 3}$:

1. Factor the numerators and denominators.

$$= \frac{(x-1)(x+3)}{(x+2)(x+3)} \cdot \frac{(x+4)(x-3)}{(x-1)(x-3)}$$

2. Cancel pairs of factors common to the numerators and denominators.

$$= \frac{\overset{1}{\cancel{(x-1)}}\overset{1}{\cancel{(x+3)}}}{(x+2)\underset{1}{\cancel{(x+3)}}} \cdot \frac{(x+4)\overset{1}{\cancel{(x-3)}}}{\underset{1}{\cancel{(x-1)}}\underset{1}{\cancel{(x-3)}}}$$

3. Multiply the numerators. Multiply the denominators.

$$= \frac{x+4}{x+2}$$

So, $\dfrac{x^2 + 2x - 3}{x^2 + 5x + 6} \cdot \dfrac{x^2 + x - 12}{x^2 - 4x + 3} = \dfrac{x+4}{x+2}$.

## Dividing Rational Expressions

You have learned how to divide rational expressions by inverting the divisor and then multiplying. Here is an example.

Find $\dfrac{2x^2 + 5x - 3}{x^2 - 6x + 5} \div \dfrac{2x^2 + x - 1}{x^2 - 4x - 5}$:

1. Invert the second fraction and change $\div$ to $\cdot$. Then multiply.

$$= \frac{2x^2 + 5x - 3}{x^2 - 6x + 5} \cdot \frac{x^2 - 4x - 5}{2x^2 + x - 1}$$

2. Factor the numerators and denominators.

$$= \frac{(2x-1)(x+3)}{(x-1)(x-5)} \cdot \frac{(x+1)(x-5)}{(2x-1)(x+1)}$$

3. Cancel pairs of factors common to the numerators and denominators.

$$= \frac{\overset{1}{\cancel{(2x-1)}}(x+3)}{(x-1)\underset{1}{\cancel{(x-5)}}} \cdot \frac{\overset{1}{\cancel{(x+1)}}\overset{1}{\cancel{(x-5)}}}{\underset{1}{\cancel{(2x-1)}}\underset{1}{\cancel{(x+1)}}}$$

4. Multiply the numerators. Multiply the denominators.

$$= \frac{x+3}{x-1}$$

So, $\dfrac{2x^2 + 5x - 3}{x^2 - 6x + 5} \div \dfrac{2x^2 + x - 1}{x^2 - 4x - 5} = \dfrac{x+3}{x-1}$.

## Simplifying a Complex Fraction

A fraction that contains other fractions is called a complex fraction. To simplify a complex fraction, start by rewriting the complex fraction as a division problem. Then invert the second fraction, and multiply.

For example, to simplify the complex fraction $\dfrac{\dfrac{x^2 - 4}{x^2 - 8x + 15}}{\dfrac{12x + 24}{3x - 15}}$:

1. Rewrite the complex fraction as a division problem.

$$= \frac{x^2 - 4}{x^2 - 8x + 15} \div \frac{12x + 24}{3x - 15}$$

2. Divide by inverting the second fraction and multiplying.

$$= \frac{x^2 - 4}{x^2 - 8x + 15} \cdot \frac{3x - 15}{12x + 24}$$

3. Factor the numerators and denominators.

$$= \frac{(x + 2)(x - 2)}{(x - 5)(x - 3)} \cdot \frac{3(x - 5)}{12(x + 2)}$$

$$= \frac{(x + 2)(x - 2)}{(x - 5)(x - 3)} \cdot \frac{3(x - 5)}{2 \cdot 2 \cdot 3(x + 2)}$$

4. Cancel pairs of factors common to the numerators and denominators.

$$= \frac{\overset{1}{\cancel{(x + 2)}}(x - 2)}{\underset{1}{\cancel{(x - 5)}}(x - 3)} \cdot \frac{\overset{1}{\cancel{3}}\overset{1}{\cancel{(x - 5)}}}{2 \cdot 2 \cdot \underset{1}{\cancel{3}}\underset{1}{\cancel{(x + 2)}}}$$

5. Multiply the numerators. Multiply the denominators.

$$= \frac{x - 2}{4(x - 3)}$$

So, $\dfrac{\dfrac{x^2 - 4}{x^2 - 8x + 15}}{\dfrac{12x + 24}{3x - 15}} = \dfrac{x - 2}{4(x - 3)}$.

## Sample Problems

1. Reduce to lowest terms: $\frac{9 - x}{x - 9}$

   ☑ a. Factor −1 out of the numerator. $= \frac{-1(-9 + x)}{x - 9}$

   ☐ b. Rewrite $(-9 + x)$ as $(x - 9)$ = ________

   ☐ c. Cancel common factors. = _____

2. Reduce to lowest terms: $\frac{x^2 - 9x + 20}{16 - x^2}$

   ☑ a. Factor the numerator and denominator. $= \frac{(x - 4)(x - 5)}{(4 - x)(4 + x)}$

   ☐ b. Recall that $\frac{x - 4}{4 - x} = -1$. So rewrite the expression. = _____ · _____

   ☐ c. Multiply. = ________

***Answers to Sample Problems***

1. *b.* $\frac{-1(x - 9)}{x - 9}$

   *c.* $-1$

2. *b.* $-1, \frac{x - 5}{4 + x}$

   *c.* $-\frac{x - 5}{4 + x}$

***Answers to Sample Problems***

a. $x + 5, x + 1$

b. $x \cdot \frac{x-4}{(x+1)}$

c. $\frac{x(x-4)}{x+1}$

b. $x + 2, x + 2$

c. $\frac{x}{(x+2)} \cdot \frac{(x+4)}{(x+2)}$

d. $\frac{x(x+4)}{(x+2)^2}$

b. $\frac{9}{2x+4}$

c. $\frac{x(x+2)}{3 \cdot x} \cdot \frac{3 \cdot 3}{2(x+2)}$

d. $\frac{1}{1} \cdot \frac{3}{2}$

e. $\frac{3}{2}$

3. Find: $\frac{x^3 + 2x^2 - 15x}{x^2 + 8x + 15} \cdot \frac{x^2 - x - 12}{x^2 - 2x - 3}$

☐ a. Factor the numerators and denominators. $= \frac{x(x+5)(x-3)}{(\_\_\_\_)(x+3)} \cdot \frac{(x+3)(x-4)}{(\_\_\_\_)(x-3)}$

☐ b. Cancel common factors. = ________

☐ c. Multiply the numerators. Multiply the denominators. = ________

4. Find: $\frac{x^2 - 3x}{x^2 + x - 2} \div \frac{x^2 - x - 6}{x^2 + 3x - 4}$

✔ a. Invert the second fraction and change ÷ to ·. $= \frac{x^2 - 3x}{x^2 + x - 2} \cdot \frac{x^2 + 3x - 4}{x^2 - x - 6}$

☐ b. Factor the numerators and denominators. $= \frac{x(x-3)}{(x-1)(\_\_\_\_)} \cdot \frac{(x-1)(x+4)}{(x-3)(\_\_\_\_)}$

☐ c. Cancel common factors. = ________

☐ d. Multiply the numerators. Multiply the denominators. = ________

5. Simplify this complex fraction: $\frac{\frac{x^2 + 2x}{3x}}{\frac{2x + 4}{9}}$

✔ a. Rewrite as a division problem. $= \frac{x^2 + 2x}{3x} \div \frac{2x + 4}{9}$

☐ b. Divide. (Invert the second fraction and multiply.) $= \frac{x^2 + 2x}{3x} \cdot$ ________

☐ c. Factor the numerators and denominators. = ________

☐ d. Cancel pairs of factors common to the numerator and denominator. = ________

☐ e. Multiply the numerators. Multiply the denominators. = ________

# ADDING AND SUBTRACTING

## Summary

### Finding the Least Common Denominator (LCD) of Two or More Rational Expressions

In order to add or subtract rational expressions with different denominators, you need to find the least common denominator (LCD) of the rational expressions. This LCD is the least common multiple (LCM) of the polynomials in the denominators. You can find the LCM of a collection of polynomials in the same way you find the LCM of a collection of numbers.

To find the LCM of a collection of polynomials:

1. Factor each polynomial.
2. List each factor the greatest number of times it appears in any factorization.
3. Find the product of these factors. This is the LCM.

For example, to find the LCM of $x^2 - 4$, $x^2 + x - 6$, and $x^2 + 6x + 9$:

1. Factor each polynomial.
   $x^2 - 4 = (x + 2)(x - 2)$
   $x^2 + x - 6 = (x + 3)(x - 2)$
   $x^2 + 6x + 9 = (x + 3)(x + 3)$

2. List each factor the greatest number of times it appears in any factorization. $(x + 2)$, $(x - 2)$, $(x + 3)$, $(x + 3)$

3. Find the product of these factors. $(x + 2)(x - 2)(x + 3)(x + 3)$

So, the LCM of $x^2 - 4$, $x^2 + x - 6$, and $x^2 + 6x + 9$ is $(x + 2)(x - 2)(x + 3)(x + 3)$.

*Remember, to find the LCM of a collection of numbers:*

1. *Find the prime factorization of each number.*

   15 → $3 \cdot 5$ 25 → $5 \cdot 5$ 18 → $2 \cdot 9$ → $2 \cdot 3 \cdot 3$

2. *List each prime factor the greatest number of times it appears in any factorization.* 2, 3, 3, 5, 5

3. *Find the product of these factors. This is the LCM.*

   $2 \cdot 3 \cdot 3 \cdot 5 \cdot 5 = 450$

*The LCM of a collection of polynomials is usually left in factored form. You don't have to do the multiplication.*

*Remember, to add fractions with different denominators:*

1. *Factor each denominator into its prime factors.*
$$\frac{5}{6} + \frac{7}{15} = \frac{5}{2 \cdot 3} + \frac{7}{3 \cdot 5}$$
2. *Find the LCD of both fractions. This is the LCM of the denominators.*
$$LCD = 2 \cdot 3 \cdot 5$$
3. *Rewrite each fraction with this LCD.*
$$= \frac{5 \cdot 5}{2 \cdot 3 \cdot 5} + \frac{7 \cdot 2}{3 \cdot 5 \cdot 2}$$
4. *Add the numerators. Keep the denominator the same.*
$$= \frac{25 + 14}{30}$$
$$= \frac{39}{30}$$
5. *Factor and reduce the fraction to lowest terms as appropriate.*
$$= \frac{\overset{1}{\cancel{3}} \cdot 13}{\underset{1}{\cancel{3}} \cdot 2 \cdot 5}$$
$$= \frac{13}{10}$$

*So,* $\frac{5}{6} + \frac{7}{15} = \frac{13}{10}$.

*In the third step you may have been tempted to cancel common factors. But this would get you back to where you started. Remember to add first, then factor and cancel common factors.*

## Adding Rational Expressions with Different Denominators

You can add rational expressions with different denominators in much the same way you add fractions with different denominators. The key idea is to write the rational expressions with the same denominator.

To add rational expressions with different denominators:

1. Factor each denominator.
2. Find the LCM of the denominators. This is the LCD of the rational expressions.
3. Rewrite each algebraic fraction with this LCD.
4. Add the numerators. The denominator stays the same.
5. Factor and reduce the rational expression to lowest terms as appropriate.

For example, to find $\frac{3}{m^3n} + \frac{5}{m^2n^2}$:

1. Factor each denominator: $= \frac{3}{m \cdot m \cdot m \cdot n} + \frac{5}{m \cdot m \cdot n \cdot n}$
2. Find the LCD of the rational expressions. $LCD = m \cdot m \cdot m \cdot n \cdot n$
3. Rewrite each algebraic fraction with the LCD, $m \cdot m \cdot m \cdot n \cdot n$. $= \frac{3 \cdot n}{m \cdot m \cdot m \cdot n \cdot n} + \frac{5 \cdot m}{m \cdot m \cdot n \cdot n \cdot m}$
4. Add the numerators. The denominator stays the same. $= \frac{3n + 5m}{m^3n^2}$

This rational expression is in lowest terms.

So, $\frac{3}{m^3n} + \frac{5}{m^2n^2} = \frac{3n + 5m}{m^3n^2}$.

As another example, to find $\frac{x+2}{x^2-4x} + \frac{x+2}{x^2-16}$:

1. Factor each denominator: $= \frac{x+2}{x(x-4)} + \frac{x+2}{(x+4)(x-4)}$
2. Find the LCD of the rational expressions. $LCD = x(x+4)(x-4)$
3. Rewrite each algebraic fraction with the LCD, $x(x-4)(x+4)$. $= \frac{(x+2)(x+4)}{x(x-4)(x+4)} + \frac{(x+2)x}{(x+4)(x-4)x}$
4. Add the numerators. The denominator stays the same. $= \frac{x^2 + 6x + 8 + x^2 + 2x}{x(x+4)(x-4)}$ $= \frac{2x^2 + 8x + 8}{x(x+4)(x-4)}$
5. Factor and reduce the rational expression to lowest terms. $= \frac{2(x+2)(x+2)}{x(x+4)(x-4)}$

So, $\frac{x+2}{x^2-4x} + \frac{x+2}{x^2-16} = \frac{2(x+2)(x+2)}{x(x+4)(x-4)}$.

## Subtracting Rational Expressions with Different Denominators

You can subtract rational expressions with different denominators in much the same way you subtract fractions with different denominators. The key idea is to write the rational expressions with the same denominator.

To subtract rational expressions with different denominators:

1. Factor each denominator.
2. Find the LCM of the denominators. This is the LCD of the rational expressions.
3. Rewrite each algebraic fraction with this LCD.
4. Subtract the numerators. The denominator stays the same.
5. Factor and reduce the rational expression to lowest terms as appropriate.

For example, to find $\frac{4}{ab} - \frac{7}{a^3b^2}$:

1. Factor each denominator: $= \frac{4}{a \cdot b} - \frac{7}{a \cdot a \cdot a \cdot b \cdot b}$
2. Find the LCD of the rational expressions. $\text{LCD} = a \cdot a \cdot a \cdot b \cdot b$
3. Rewrite each algebraic fraction with the LCD, $a \cdot a \cdot a \cdot b \cdot b$. $= \frac{4 \cdot a \cdot a \cdot b}{a \cdot b \cdot a \cdot a \cdot b} - \frac{7}{a \cdot a \cdot a \cdot b \cdot b}$
4. Subtract the numerators. The denominator stays the same. $= \frac{4a^2b - 7}{a^3b^2}$

This rational expression is in lowest terms.

So, $\frac{4}{ab} - \frac{7}{a^3b^2} = \frac{4a^2b - 7}{a^3b^2}$.

As another example, to find $\frac{10}{x^2 + x - 6} - \frac{3}{x - 2}$:

1. Factor each denominator: $= \frac{10}{(x - 2)(x + 3)} - \frac{3}{x - 2}$
2. Find the LCD of the rational expressions. $\text{LCD} = (x - 2)(x + 3)$
3. Rewrite each algebraic fraction with the LCD $(x - 2)(x + 3)$. $= \frac{10}{(x - 2)(x + 3)} - \frac{3(x + 3)}{(x - 2)(x + 3)}$
4. Subtract the numerators. The denominator stays the same. $= \frac{10 - 3(x + 3)}{(x - 2)(x + 3)}$

$$= \frac{10 - 3x - 9}{(x - 2)(x + 3)}$$

$$= \frac{1 - 3x}{(x - 2)(x + 3)}$$

This rational expression is in lowest terms.

So, $\frac{10}{x^2 + x - 6} - \frac{3}{x - 2} = \frac{1 - 3x}{(x - 2)(x + 3)}$.

*Remember, to subtract fractions with different denominators:*

1. *Factor each denominator into its prime factors.*
   $\frac{5}{6} - \frac{7}{15} = \frac{5}{2 \cdot 3} - \frac{7}{3 \cdot 5}$
2. *Find the LCD of both fractions. This is the LCM of the denominators.*
   $LCD = 2 \cdot 3 \cdot 5$
3. *Rewrite each fraction with this LCD.*
   $= \frac{5 \cdot 5}{2 \cdot 3 \cdot 5} - \frac{7 \cdot 2}{3 \cdot 5 \cdot 2}$
4. *Subtract the numerators. Keep the denominator the same.*
   $= \frac{25 - 14}{30}$
   $= \frac{11}{30}$

*So,* $\frac{5}{6} - \frac{7}{15} = \frac{11}{30}$.

*In the third step you may have been tempted to cancel common factors. But this would get you back to where you started. Remember to subtract first, then factor and cancel common factors.*

## Simplifying a Complex Fraction

Recall that a fraction that contains other fractions is called a complex fraction. Here's one way to simplify certain complex fractions:

1. Perform any addition or subtraction in the numerator or denominator.
2. Rewrite the complex fraction as a division problem.
3. Divide.

For example, to find $\dfrac{\frac{3}{x+1}+5}{\frac{3}{x+1}}$:

1. Perform the addition in the numerator. The least common denominator of $\frac{3}{x+1}$ and 5 is $(x + 1)$.

$$= \dfrac{\frac{3}{x+1}+\frac{5(x+1)}{x+1}}{\frac{3}{x+1}}$$

$$= \dfrac{\frac{3+5(x+1)}{x+1}}{\frac{3}{x+1}}$$

$$= \dfrac{\frac{3+5x+5}{x+1}}{\frac{3}{x+1}}$$

$$= \dfrac{\frac{8+5x}{x+1}}{\frac{3}{x+1}}$$

2. Rewrite the complex fraction as a division problem.

$$= \frac{8+5x}{x+1} \div \frac{3}{x+1}$$

3. To divide, invert the second fraction and replace $\div$ with $\cdot$.

$$= \frac{8+5x}{x+1} \cdot \frac{x+1}{3}$$

4. Cancel common factors.

$$= \frac{8+5x}{3}$$

So, $\dfrac{\frac{3}{x+1}+5}{\frac{3}{x+1}} = \dfrac{8+5x}{3}$.

## Sample Problems

1. Find the LCM of $x^2 + 7x + 12$, $x^2 - 9$, and $x^2 + 8x + 16$.

☐ a. Factor each polynomial.

$x^2 + 7x + 12 = (\_\_\_\_)(\_\_\_\_)$
$x^2 - 9 = (\_\_\_\_)(\_\_\_\_)$
$x^2 + 8x + 16 = (\_\_\_\_)(\_\_\_\_)$

☐ b. List each factor the greatest number of times it appears in any factorization.

____, ____, ____, ____

☐ c. Find the product of the factors in the list. This is the LCM.

$(\_\_\_\_)(\_\_\_\_)(\_\_\_\_)(\_\_\_\_)$

2. Find: $\frac{2}{x^2 + 2x - 15} + \frac{x}{x^2 - 9}$

✓ a. Factor each denominator.

$= \frac{2}{(x+5)(x-3)} + \frac{x}{(x+3)(x-3)}$

☐ b. Find the LCD of the rational expressions.

LCD $= (\_\_\_\_)(\_\_\_\_)(\_\_\_\_)$

☐ c. Rewrite each rational expression using this LCD.

$= \frac{2(\_\_\_\_)}{(x+5)(x-3)(\_\_\_\_)} + \frac{x(\_\_\_\_)}{(x+3)(x-3)(\_\_\_\_)}$

☐ d. Add the numerators. The denominator stays the same.

$= \frac{2(\_\_\_\_) + x(\_\_\_\_)}{(x+5)(x+3)(x-3)}$

$= \frac{2x + 6 + x^2 + 5x}{(x+5)(x+3)(x-3)}$

$= \_\_\_\_\_\_\_\_$

3. Find: $\frac{x}{x^2 - 6x + 9} - \frac{3}{x^2 - 9}$

✓ a. Factor each denominator.

$= \frac{x}{(x-3)(x-3)} - \frac{3}{(x+3)(x-3)}$

☐ b. Find the LCD of the rational expressions.

LCD $= (\_\_\_\_)(\_\_\_\_)(\_\_\_\_)$

☐ c. Rewrite each rational expression using this LCD.

$= \frac{x(\_\_\_\_)}{(x-3)(x-3)(\_\_\_\_)} - \frac{3(\_\_\_\_)}{(x+3)(x-3)(\_\_\_\_)}$

☐ d. Subtract the numerators. The denominator stays the same.

$= \frac{x(\_\_\_\_) - 3(\_\_\_\_)}{\_\_\_\_\_\_}$

$= \_\_\_\_\_\_$

### *Answers to Sample Problems*

*1. a. $x + 3, x + 4$ (in either order)*
*$x + 3, x - 3$ (in either order)*
*$x + 4, x + 4$ (in either order)*

*b. $x + 3, x - 3, x + 4, x + 4$ (in any order)*

*c. $(x + 3)(x - 3)(x + 4)(x + 4)$ (in any order)*

*2. b. $x + 5, x + 3, x - 3$ (in any order)*

*c. top: $x + 3, x + 5$*
*bottom: $x + 3, x + 5$*

*d. $x + 3, x + 5$*

$\frac{x^2 + 7x + 6}{(x+5)(x+3)(x-3)}$ *or* $\frac{(x+1)(x+6)}{(x+5)(x+3)(x-3)}$

*3. b. $x - 3, x - 3, x + 3$ (in any order)*

*c. top: $x + 3, x - 3$*
*bottom: $x + 3, x - 3$*

*d. top: $x + 3, x - 3$*
*bottom: $(x - 3)(x - 3)(x + 3)$*

$\frac{x^2 + 3x - 3x + 9}{(x-3)(x-3)(x+3)}$ *or* $\frac{x^2 + 9}{(x-3)(x-3)(x+3)}$

**Answers to Sample Problems**

b. $z + 5, z - 5$

c. top: $z + 5, z - 5$
bottom: $z + 5, z - 5$

d. top: $z + 5, -1, z - 5$
bottom: $(z + 5)(z - 5)$

$\frac{2z + 10 - z + 5 - 10}{(z + 5)(z - 5)}$

$\frac{z + 5}{(z + 5)(z - 5)}$

e. $\frac{1}{z - 5}$

a. $xy^2$

b. $\frac{2y}{xy^2}, \frac{1}{xy^2}$

c. $\frac{2y + 1}{xy^2}$

d. $3 \div \frac{2y + 1}{xy^2}$

e. $\frac{3xy^2}{2y + 1}$

4. Find: $\frac{2}{z - 5} + \frac{-1}{z + 5} - \frac{10}{z^2 - 25}$

- ☑ a. Factor each denominator. $= \frac{2}{z - 5} + \frac{-1}{z + 5} - \frac{10}{(z + 5)(z - 5)}$
- ☐ b. Find the LCD of the rational expressions. LCD = (_____)(_____)
- ☐ c. Rewrite each rational expression using this LCD. $= \frac{2(\_\_\_)}{(z - 5)(\_\_\_)} + \frac{-1(\_\_\_)}{(z + 5)(\_\_\_)} - \frac{10}{(z + 5)(z - 5)}$
- ☐ d. Add or subtract the numerators. The denominator stays the same. $= \frac{2(\_\_\_) + (\_)(\_\_\_) - 10}{\_\_\_\_\_}$

  = __________

  = __________
- ☐ e. Factor and reduce the rational expression to lowest terms. = __________

5. Simplify this complex fraction: $\frac{3}{\frac{2}{xy} + \frac{1}{xy^2}}$

- ☐ a. To add the fractions in the denominator, find the LCD. LCD = _____
- ☐ b. Rewrite each fraction using this LCD. $= \frac{3}{\_\_\_ + \_\_\_}$
- ☐ c. Add the fractions in the denominator. $= \frac{3}{\_\_\_}$
- ☐ d. Rewrite the complex fraction as a division problem. = __________
- ☐ e. Divide. = __________

# HOMEWORK

## Homework Problems

Circle the homework problems assigned to you by the computer, then complete them below.

### Explain
### Negative Exponents

1. Find: $5^{-3}$

2. Find: $7^{-2} \cdot 7^{3}$

3. Find: $\dfrac{50}{10^{-2} + 5^{-2}}$

In problems 4 through 9, simplify each expression. Use only positive exponents in your answers.

4. $\left(\dfrac{2m}{n}\right)^{-3}$

5. $\dfrac{(2m)^{-3}}{n^3}$

6. $(n^4p^{-3})^5$

7. $(2s^3t)^{-5} \cdot (2st^3)^6$

8. $\dfrac{(3uv^4)(9u^3v)^{-2}}{(3u^5v^{-4})^{-3}}$

9. All matter is made up of tiny particles called atoms. Through experimentation, it has been found that the diameters of atoms range from $1 \times 10^{-8}$ cm to $5 \times 10^{-8}$ cm. Rewrite each of these numbers in expanded form.

10. The monthly payment $E$ on a loan of amount $P$ can be computed by using the formula below, where $r$ is the monthly interest rate, and $n$ is the number of months for which the loan is made. Find the monthly payment on a \$15,000 loan for 4 years (48 months) if the monthly interest rate is 1%.

$$E = \frac{Pr}{1 - (1 + r)^{-n}}$$

11. Write $\left(\dfrac{3x^4y}{z^5}\right)^{-2}$ using only positive exponents.

12. Suppose $x = 3$ and $y = 5$.

    a. Is $x^{-1}y^{-1} = \dfrac{1}{x} \cdot \dfrac{1}{y}$?

    b. Is $x^{-1}y^{-1} = \dfrac{1}{xy}$?

    c. Is $x^{-1} + y^{-1} = \dfrac{1}{x} + \dfrac{1}{y}$?

    d. Is $x^{-1} + y^{-1} = \dfrac{1}{x + y}$?

### Multiplying and Dividing

13. Reduce to lowest terms: $\dfrac{y - 13}{13 - y}$

14. Reduce to lowest terms: $\dfrac{y^2 + y - 20}{16 - y^2}$

15. Find: $\dfrac{6 - x}{x + 7} \cdot \dfrac{x^2 + 7x}{x - 6}$

16. $\dfrac{a^2 - 5a - 14}{a^2 - 2a - 35} \cdot \dfrac{a^2 + 6a + 5}{a^2 - a - 6}$

17. Find: $\dfrac{3x^2 - 7x + 2}{x^2 + 4x - 12} \cdot \dfrac{2x^2 + 12x}{3x^2 + 2x - 1}$

18. Find: $\dfrac{(x + 3)(x - 2)}{x^2 - 4} \div \dfrac{(x + 3)(x - 4)}{x^2 - 16}$

19. Find: $\dfrac{y^2 + 15y + 56}{8y^2 - 10y + 3} \div \dfrac{y + 7}{4y - 3}$

20. Find: $\dfrac{x(x - 9)}{x^2 + 3x + 2} \div \dfrac{x^2 - 81}{x(x + 2)}$

21. Find: $\dfrac{z^2 + 8z + 15}{z^2 - 16} \div \dfrac{z^2 + 10z + 25}{z + 4}$

In Problems 22, 23 and 24, simplify the complex fractions. Reduce your answer to lowest terms.

22. $\dfrac{\dfrac{x^2 - 13x + 42}{x^2 - 4}}{\dfrac{3x - 18}{3x - 6}}$

23. $\dfrac{\dfrac{2z^2 - 4z}{3z^3 + 6z^2}}{\dfrac{z - 2}{z + 2}}$

24. $\dfrac{\dfrac{2w^2 - 4w}{2w^2 - 2}}{\dfrac{w^2 - 5w + 6}{w^2 - 2w - 3}}$

## Adding and Subtracting

25. Find the LCM of $x^2 + 10x + 25$ and $x^2 - 3x - 40$.

26. Find the LCM of $x$, $x^2 - 16$, and $x^2 + 7x + 12$.

27. Find: $\frac{5}{xy^3} + \frac{14}{x^2y^2}$

28. Find: $\frac{-1}{x^2 + 3x + 2} + \frac{2}{x^2 + 2x}$

29. Find: $\frac{4}{xy^2z} - \frac{3}{xyz^2}$

30. Find: $\frac{8}{x^2 + 14x + 49} - \frac{4}{x^2 - 49}$

31. Find: $\frac{-1}{y+2} + \frac{2}{y-2} - \frac{4}{y^2 - 4}$

32. Find: $\frac{1}{x^2 + 8x} - \frac{3}{x^2 - 64} + \frac{2}{x^2 + 6x - 16}$

33. Simplify the left side of the equation below to show it equals $\frac{1}{n}$. Then use the equation to find two fractions with 1 in the numerator whose difference is $\frac{1}{5}$. (Hint: let $n = 5$).

$$\frac{1}{n-1} - \frac{1}{n(n-1)} = \frac{1}{n}$$

34. Optometrists use the formula below to find the strength to be used for the lenses of glasses. Simplify the right side of this formula, then find the value of $P$ that corresponds to $a = 12$ and $b = 0.3$.

$$P = \frac{1}{a} + \frac{1}{b}$$

35. The total resistance, $R$, of a circuit that consists of two resistors connected in parallel with resistance $R_1$ and $R_2$ is given by the formula below. Simplify this formula, then find the resistance, $R$, if $R_1 = 3$ ohms and $R_2 = 4$ ohms.

$$R = \frac{1}{\frac{1}{R_1} + \frac{1}{R_2}}$$

36. Simplify this complex fraction: $\frac{1}{\frac{1}{a} + \frac{1}{ab^2}}$

# APPLY

## Practice Problems

Here are some additional practice problems for you to try.

### Negative Exponents

1. Find: $2^{-3}$
2. Find: $4^{-2}$
3. Find: $3^{-4}$
4. Find: $5^{-2}$
5. Find: $\frac{1}{3^{-4}}$
6. Find: $\frac{1}{5^{-3}}$
7. Find: $4^{-7} \cdot 4^{5}$
8. Find: $2^{8} \cdot 2^{-5}$
9. Find: $5^{-9} \cdot 5^{6}$
10. Find: $\frac{1}{3^{-2} + 2^{-3}}$
11. Find: $\frac{3}{4^{-3} + 5^{-2}}$
12. Find: $\frac{2}{4^{-2} + 7^{-1}}$
13. Rewrite using only positive exponents: $(a^4b^6)^{-1}$
14. Rewrite using only positive exponents: $(x^3y^5)^{-2}$
15. Rewrite using only positive exponents: $(m^6n^3p)^{-4}$
16. Rewrite using only positive exponents: $(x^{-5}b^2)^5$
17. Rewrite using only positive exponents: $(a^{-3}b^7)^4$
18. Rewrite using only positive exponents: $(x^{-6}yz^{-3})^5$
19. Rewrite using only positive exponents: $\left(\frac{c^4d^{-5}}{a^2}\right)^{-2}$
20. Rewrite using only positive exponents: $\left(\frac{x^3w^{-2}}{y^4}\right)^{-3}$
21. Rewrite using only positive exponents: $\left(\frac{m^{-4}n^5}{p^{-3}}\right)^{-4}$
22. Rewrite using only positive exponents: $(3x^3y)^{-2} \cdot (3x^2y^{-1}z^3)^4$
23. Rewrite using only positive exponents: $(4a^4b^2)^4 \cdot (4a^{-3}bc^2)^{-3}$
24. Rewrite using only positive exponents: $(5x^{-3}y^{-1}z)^{-3} \cdot (5xz^{-5})^4$
25. Write in scientific notation: 0.000057
26. Write in scientific notation: 148,000,000
27. The following number is written in scientific notation. Write it in expanded form: $4.3 \times 10^6$
28. The following number is written in scientific notation. Write it in expanded form: $1.785 \times 10^{-4}$

### Multiplying and Dividing

29. Reduce to lowest terms: $\frac{x-5}{5-x}$
30. Reduce to lowest terms: $\frac{x-3}{3-x}$
31. Reduce to lowest terms: $\frac{x^2-2x-35}{x^2-25}$
32. Reduce to lowest terms: $\frac{x^2+2x-24}{16-x^2}$
33. Reduce to lowest terms: $\frac{x^2-8x+7}{49-x^2}$
34. Reduce to lowest terms: $\frac{x^2-11x+30}{x^2-36}$
35. Reduce to lowest terms: $\frac{x^2-8x-9}{81-x^2}$
36. Find: $\frac{x^2+9x+14}{x^2+2x-15} \cdot \frac{x^2-4x+3}{x^2+6x-7}$
37. Find: $\frac{x^2-7x+12}{x^2+2x-15} \cdot \frac{x^2-x-30}{x^2-3x-18}$
38. Find: $\frac{x^2+5x+6}{x^2-5x-6} \cdot \frac{x^2-10x+24}{x^2-x-12}$
39. Find: $\frac{3x^2-6x}{x+1} \cdot \frac{x-1}{2-x}$
40. Find: $\frac{5x^2-25x}{x-3} \cdot \frac{x+3}{5-x}$

41. Find: $\frac{2x^2 - 6x}{x - 5} \cdot \frac{x + 5}{3 - x}$

42. Find: $\frac{x^2 + 2x - 15}{x^2 - 25} \cdot \frac{x^2 + 3x + 2}{x^2 - 2x - 3}$

43. Find: $\frac{x^2 + 4x - 45}{x^2 - 81} \cdot \frac{x^2 - 4x - 45}{x^2 - 7x + 10}$

44. Find: $\frac{x^2 + 10x + 21}{x^2 - 2x - 15} \cdot \frac{x^2 - x - 20}{x^2 - 16}$

45. Find: $\frac{x^2 + 2x - 35}{x^2 + x - 90} \div \frac{x^2 + 10x + 21}{x^2 + x - 90}$

46. Find: $\frac{x^2 + 2x - 35}{x^2 + x - 90} \div \frac{x^2 + 10x + 21}{x^2 + x - 90}$

47. Find: $\frac{x^2 - 2x - 3}{x^2 + 4x - 5} \div \frac{x^2 - 10x + 21}{x^2 + 4x - 5}$

48. Find: $\frac{5x - 25}{x^2 - 49} \div \frac{x^2 - 9x + 20}{x^2 - 11x + 28}$

49. Find: $\frac{x^2 + 7x + 6}{x^2 - 5x - 6} \div \frac{3x + 18}{x^2 + 5x - 66}$

50. Find: $\frac{4x - 16}{x^2 - 36} \div \frac{x^2 - x - 12}{x^2 + 9x + 18}$

51. Simplify the complex fraction below. Write your answer in lowest terms.

$$\frac{\frac{x^2 + 3x - 70}{x^2 - 49}}{\frac{x^2 + 9x - 10}{3x^2 - 3x}}$$

52. Simplify the complex fraction below. Write your answer in lowest terms.

$$\frac{\frac{x^2 - 4x - 45}{4x^2 + 20x}}{\frac{x^2 + 2x - 99}{x^2 - 121}}$$

53. Simplify the complex fraction below. Write your answer in lowest terms.

$$\frac{\frac{x^2 - 8x - 33}{x^2 - 9}}{\frac{x^2 - 9x - 22}{5x^2 + 10x}}$$

54. Simplify the complex fraction below. Write your answer in lowest terms.

$$\frac{\frac{x^2 - 7x + 6}{x^2 + 8x + 12}}{\frac{3x^2 - 3x}{x^2 - 36}}$$

55. Simplify the complex fraction below. Write your answer in lowest terms.

$$\frac{\frac{x^2 + 2x - 15}{5x^2 + 15x}}{\frac{x^2 - 2x - 35}{x^2 - 9}}$$

56. Simplify the complex fraction below. Write your answer in lowest terms.

$$\frac{\frac{x^2 + x - 20}{x^2 + 5x + 4}}{\frac{2x^2 + 10x}{x^2 - 16}}$$

## Adding and Subtracting

57. Find the LCM of $x^2 + 7x + 12$ and $x^2 - 3x - 28$.

58. Find the LCM of $x^2 + 11x + 28$ and $x^2 + 2x - 8$.

59. Find the LCM of $x^2 + 4x$, $x^2 + 3x - 4$, and $x^2 - 2x + 1$.

60. Find the LCM of $x^2 - 6x$, $x^2 - 5x - 6$, and $x^2 + 2x + 1$.

61. Find: $\frac{4}{9a} + \frac{2}{3b}$

62. Find: $\frac{3}{8m} + \frac{7}{10n}$

63. Find: $\frac{2}{3x} + \frac{2}{6y}$

64. Find: $\frac{4}{x^2 - 4x - 12} + \frac{3}{x - 6}$

65. Find: $\frac{2}{x + 10} + \frac{5}{x^2 + 5x - 50}$

66. Find: $\frac{5}{x^2 + 4x - 21} + \frac{1}{x + 7}$

67. Find: $\frac{x}{x + 1} + \frac{x + 4}{x - 4}$

68. Find: $\frac{3x}{x + 3} + \frac{x + 2}{x - 7}$

69. Find: $\frac{5x}{x - 3} + \frac{x + 1}{x + 2}$

70. Find: $\frac{x + 2}{x^2 + 7x + 12} + \frac{x - 1}{x^2 + x - 12}$

71. Find: $\frac{x + 4}{x^2 + 5x + 6} + \frac{x - 3}{x^2 - 3x - 10}$

72. Find: $\frac{x - 5}{x^2 - 6x + 8} + \frac{x + 3}{x^2 + 2x - 8}$

73. Find: $\frac{4}{m^2n} - \frac{1}{mn^2}$

74. Find: $\frac{5}{abc} - \frac{7}{b^2}$

75. Find: $\frac{3}{xyz} - \frac{2}{x^2}$

76. Find: $\frac{7x}{x + 6} - \frac{5}{x - 1}$

77. Find: $\frac{3x}{x-8} - \frac{11}{x+3}$

78. Find: $\frac{8x}{x-7} - \frac{3}{x+1}$

79. Find: $\frac{x+2}{x^2+6x+5} - \frac{x+3}{x^2+4x-5}$

80. Find: $\frac{x+5}{x^2-8x+12} - \frac{x-1}{x^2-3x-18}$

81. Find: $\frac{x+4}{x^2+x-2} - \frac{x+1}{x^2+2x-3}$

82. Simplify this complex fraction: $\dfrac{\frac{4}{x}+\frac{1}{y}}{\frac{3}{x}-\frac{5}{y}}$

83. Simplify this complex fraction: $\dfrac{\frac{1}{x+1}+\frac{2}{x}}{\frac{4}{x}-\frac{3}{x+1}}$

84. Simplify this complex fraction: $\dfrac{\frac{7}{x}-\frac{3}{y}}{\frac{1}{x}+\frac{2}{y}}$

# EVALUATE

## Practice Test

Take this practice test to be sure that you are prepared for the final quiz in Evaluate.

1. Find: $\left(\frac{1}{2}\right)^{-3}$
2. Fill in the blanks below by writing the numbers in either scientific notation or expanded form.

$$73901 = \_\_\_\_\_\_ \times 10^4$$

$$0.00004003 = 4.003 \times 10^{\_\_}$$

$$2.081 \times 10^2 = \_\_\_\_\_\_\_\_\_\_$$

$$9{,}019 \times 10^{-5} = \_\_\_\_\_\_\_\_\_\_$$

3. Rewrite $\frac{a^{-8}b^{13}}{a^{-2}b^{-2}}$ using only positive exponents.
4. Rewrite $\frac{(2xy^5)(8x^2y)^{-3}}{(2x^4y^{-5})^{-2}}$ using only positive exponents.
5. Reduce to lowest terms: $\frac{2x^3 - 8x}{4x - 2x^2}$
6. Multiply and reduce to lowest terms:

   a. $\frac{x+3}{x^2-4} \cdot \frac{x^2 - 10x + 16}{x-8}$

   b. $\frac{x^2 + x - 2}{x^3 - 6x^2} \cdot \frac{2x^2 - 14x + 12}{x+2}$

7. Divide and reduce your result to lowest terms:

   a. $\frac{3x^2 - 75}{x^2 - 10x + 25} \div \frac{-x-5}{x-5}$

   b. $\frac{y^2 + 5y + 4}{y^2 + y - 30} \div \frac{y^2 + 2y + 1}{y-5}$

8. Simplify this complex fraction: $\dfrac{\frac{4x+12}{5x-5}}{\frac{2x^2-18}{x^2-2x+1}}$
9. Find the LCM of $x$, $x^2 + 12x + 35$, and $x^2 - 25$.
10. Add and reduce your answer to lowest terms.

$$\frac{4}{b^2 + 4b + 3} + \frac{3}{b^2 + 3b + 2}$$

11. Subtract and reduce your answer to lowest terms.

$$\frac{3y}{y^2 + 7y + 10} - \frac{2y}{y^2 + 6y + 8}$$

12. Simplify this complex fraction: $\dfrac{\frac{3}{5}}{\frac{1}{x} + 3}$

# ANSWERS

## Homework

**1.** $\frac{1}{125}$ or $\frac{1}{5^3}$ **3.** 1,000 **5.** $\frac{1}{8m^3n^3}$ **7.** $\frac{2t^{13}}{s^9}$

**9.** 0.00000001 cm, 0.00000005 cm **11.** $\frac{z^{10}}{9x^8y^2}$

**13.** –1 **15.** $-x$ **17.** $\frac{2x}{x+1}$ **19.** $\frac{y+8}{2y-1}$

**21.** $\frac{z+3}{(z-4)(z+5)}$ **23.** $\frac{2}{3z}$ **25.** $(x+5)(x+5)(x-8)$

**27.** $\frac{5x+14y}{x^2y^3}$ **29.** $\frac{4z-3y}{xy^2z^2}$ **31.** $\frac{1}{y-2}$

**33.** $\frac{1}{n-1} - \frac{1}{n(n-1)}$

$= \frac{n}{n(n-1)} - \frac{1}{n(n-1)}$

$= \frac{n-1}{n(n-1)}$

$= \frac{1}{n}$

The two fractions are $\frac{1}{4}$ and $\frac{1}{20}$:

$\frac{1}{4} - \frac{1}{20} = \frac{1}{5}$

**35.** $R = \frac{R_1R_2}{R_1 + R_2}$; $R = \frac{12}{7}$ ohms

**39.** $-\frac{3x(x-1)}{x+1}$ **41.** $\frac{-2x(x+5)}{x-5}$ or $\frac{2x(x+5)}{5-x}$ **43.** $\frac{x+5}{x-2}$

**45.** $\frac{x-5}{x+3}$ **47.** $\frac{x+1}{x-7}$ **49.** $\frac{x+11}{3}$ **51.** $\frac{3x}{x+7}$ **53.** $\frac{5x}{x-3}$

**55.** $\frac{(x-3)(x-3)}{5x(x-7)}$ or $\frac{x^2-6x+9}{5x^2-35x}$ **57.** $(x+3)(x+4)(x-7)$

**59.** $x(x+4)(x-1)(x-1)$ **61.** $\frac{4b+6a}{9ab}$ **63.** $\frac{2y+x}{3xy}$

**65.** $\frac{2x-5}{(x+10)(x-5)}$ **67.** $\frac{2x^2+x+4}{(x+1)(x-4)}$ **69.** $\frac{6x^2+8x-3}{(x-3)(x+2)}$

**71.** $\frac{2x^2-x-29}{(x+2)(x+3)(x-5)}$ **73.** $\frac{4n-m}{m^2n^2}$ **75.** $\frac{3x-2yz}{x^2yz}$

**77.** $\frac{3x^2-2x+88}{(x-8)(x+3)}$ **79.** $\frac{-3x-5}{(x+5)(x+1)(x-1)}$

**81.** $\frac{4x+10}{(x+2)(x+3)(x-1)}$ **83.** $\frac{3x+2}{x+4}$

## Practice Problems

**1.** $\frac{1}{8}$ **3.** $\frac{1}{81}$ **5.** 81 **7.** $\frac{1}{16}$ or $4^{-2}$ **9.** $\frac{1}{125}$ or $5^{-3}$ **11.** $\frac{4800}{89}$

**13.** $\frac{1}{a^4b^6}$ **15.** $\frac{1}{m^{24}n^{12}p^4}$ **17.** $\frac{b^{28}}{a^{12}}$ **19.** $\frac{a^4d^{10}}{c^8}$ **21.** $\frac{m^{16}}{n^{20}p^{12}}$

**23.** $\frac{4a^{25}b^5}{c^6}$ **25.** $5.7 \times 10^{-5}$ **27.** 4,300,000 **29.** –1

**31.** $\frac{x-7}{x-5}$ **33.** $-\frac{x-1}{7+x}$ **35.** $-\frac{x+1}{9+x}$ **37.** $\frac{x-4}{x+3}$

## Practice Test

**1.** 8 **2.** 7.3901, $10^{-5}$, 208.1, and 0.00009019

**3.** $\frac{b^{15}}{a^6}$ **4.** $\frac{x^3}{64y^8}$ **5.** $-x-2$ **6a.** $\frac{x+3}{x+2}$ **b.** $\frac{2(x-1)^2}{x^2}$

**7a.** –3 **b.** $\frac{y+4}{(y+6)(y+1)}$ **8.** $\frac{2(x-1)}{5(x-3)}$

**9.** LCM = $x(x+5)(x-5)(x+7)$

**10.** $\frac{7b+17}{(b+1)(b+2)(b+3)}$ **11.** $\frac{y}{(y+4)(y+5)}$ **12.** $\frac{3x}{5(1+3x)}$

# LESSON 8.3 – EQUATIONS WITH FRACTIONS

# OVERVIEW

***Here is what you'll learn in this lesson:***

***Solving Equations***

*a. Solving equations with rational expressions*

*b. Solving for an unknown in a formula involving a rational expression*

Suppose you want to figure out a baseball pitcher's earned run average, or estimate the population of fish in a lake, or figure out how tall a building is based on a scale model. For each of these examples, you can figure out the answer by setting up an equation that involves fractions or ratios.

However, even with all the techniques you have for solving equations, solving an equation with fractions can be tricky. There may not be any solution at all, or the solution you find might not check when you plug it back into the original equation.

In this lesson you will learn how to solve equations that have fractions in them, and you will learn how to identify extraneous, or false solutions.

# EXPLAIN

## SOLVING EQUATIONS

### Summary

#### Solving Equations with Rational Expressions

When you solve an equation that contains a rational expression, it helps to clear the fraction in the equation. To do this, multiply both sides of the equation by the least common denominator (LCD) of the fractions.

To solve an equation that contains rational expressions:

1. Clear the fractions by multiplying both sides of the equation by the LCD of the fractions.
2. Distribute the LCD and simplify.
3. Finish solving for the variable.

For example, to solve $2x - \frac{1}{3} = \frac{2x}{3} + \frac{2x}{5} - \frac{3}{15}$ for $x$:

1. Multiply by the LCD of the fractions, 15. $\quad 15 \cdot (2x - \frac{1}{3}) = 15 \cdot \left(\frac{2x}{3} + \frac{2x}{5} - \frac{3}{15}\right)$
2. Distribute the LCD and simplify. $\quad 15 \cdot 2x - 15 \cdot \frac{1}{3} = 15 \cdot \frac{2x}{3} + 15 \cdot \frac{2x}{5} - 15 \cdot \frac{3}{15}$

$$30x - 5 = 10x + 6x - 3$$

$$30x - 5 = 16x - 3$$

3. Finish solving for $x$.

$$14x = 2$$

$$x = \frac{2}{14}$$

$$x = \frac{1}{7}$$

*Here's how to find the LCD of $\frac{1}{3}$, $\frac{2x}{5}$, and $\frac{3}{15}$:*

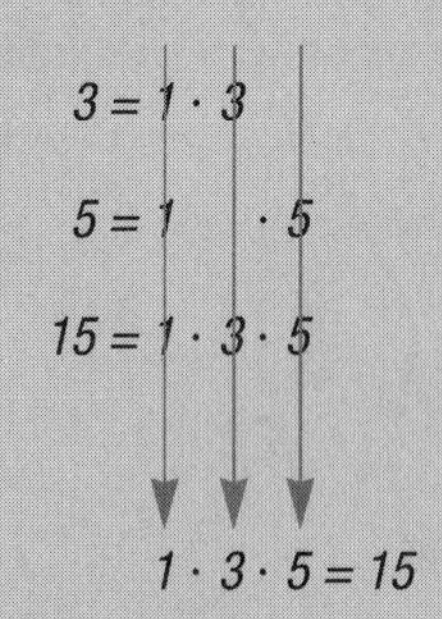


#### Checking for Extraneous Solutions

If an equation contains a fraction with a variable in the denominator, the solution of the equation might be extraneous (false).

To check for an extraneous solution:

1. Solve the equation for the variable.
2. Substitute the solution into the original equation and simplify.
3. Look at the denominators of the fractions. If any denominator is zero, the solution is extraneous.

For example, to determine if $\frac{4}{x+3} = \frac{1}{x} - \frac{12}{x(x+3)}$ has an extraneous solution:

The LCD of $\frac{4}{x+3}$, $\frac{1}{x}$, and $\frac{12}{x(x+3)}$ is $x(x+3)$.

1. Solve the equation for $x$.

$$\frac{4}{x+3} = \frac{1}{x} - \frac{12}{x(x+3)}$$

$$x(x+3) \cdot \frac{4}{x+3} = x(x+3) \cdot \left(\frac{1}{x} - \frac{12}{x(x+3)}\right)$$

$$x(x+3) \cdot \frac{4}{x+3} = x(x+3) \cdot \frac{1}{x} - x(x+3) \cdot \frac{12}{x(x+3)}$$

$$4x = x+3 - 12$$

$$3x = -9$$

$$x = -3$$

2. Substitute $-3$ for $x$ in the original equation.

$$\frac{4}{\mathbf{-3}+3} = \frac{1}{\mathbf{-3}} - \frac{12}{\mathbf{-3}(\mathbf{-3}+3)}$$

3. Check the denominators. When you substitute $x = -3$, two of the fractions have a denominator of zero.

$$\frac{4}{0} = \frac{1}{-3} - \frac{12}{0}$$

So, the solution is extraneous. This equation has no solution.

## Using Cross Multiplication to Solve Proportions

*Proportions have exactly one term on each side of the equation. Here are some examples of proportions:*

$$\frac{5}{x} = \frac{10}{20} \qquad \frac{3x-2}{3} = \frac{1}{2} \qquad \frac{-5y}{30} = 5$$

An equation that sets one fraction equal to another fraction is called a proportion. An easy way to solve a proportion is to "cross multiply."

To solve a proportion using cross multiplication:

1. Multiply the numerator of the first fraction by the denominator of the second fraction.
2. Multiply the denominator of the first fraction by the numerator of the second fraction.
3. Set the two products equal to each other.
4. Finish solving for the variable.

For example, to solve the proportion $\frac{x-1}{2} = \frac{26}{4}$:

1. Multiply the numerator of $\frac{\mathbf{x-1}}{2}$ by the denominator of $\frac{26}{\mathbf{4}}$.

$$(x-1) \cdot 4$$

2. Multiply the denominator of $\frac{x-1}{\mathbf{2}}$ by the numerator of $\frac{\mathbf{26}}{4}$.

$$2 \cdot 26$$

3. Set the products equal to each other. $2 \cdot 26 = (x-1) \cdot 4$

4. Finish solving for $x$.

$$52 = 4x - 4$$

$$56 = 4x$$

$$14 = x$$

## Sample Problems

1. Solve the equation $\frac{x-2}{4} + 3 = \frac{5}{6} - \frac{x}{6}$ for $x$. Determine if the solution is extraneous.

☐ a. Multiply both sides by the LCD of the fractions. $\_\_\_\left(\frac{x-2}{4} + 3\right) = \_\_\_\left(\frac{5}{6} - \frac{x}{6}\right)$

☐ b. Distribute the LCD and simplify. $\_\_\_\left(\frac{x-2}{4}\right) + \_\_\_ \cdot 3 = \_\_\_ \cdot \frac{5}{6} - \_\_\_ \cdot \frac{x}{6}$

$\_\_\_ + \_\_\_ = 10 - 2x$

☐ c. Finish solving for $x$. $\_\_\_ = \_\_\_$

$\_\_\_ = \_\_\_$

$x = \_\_\_$

☐ d. Substitute the solution for $x$ in the original equation. $\_\_\_ = \_\_\_$

☐ e. Is the solution extraneous? ___

2. Solve the equation $\frac{2}{x} - \frac{1}{x+2} = \frac{2}{x(x+2)}$ for $x$. Determine if the solution is extraneous.

☑ a. Multiply both sides by the LCD of the fractions. $x(x+2)\left(\frac{2}{x} - \frac{1}{x+2}\right) = x(x+2) \cdot \frac{2}{x(x+2)}$

☐ b. Distribute the LCD and simplify. $x(x+2)\left(\frac{2}{x}\right) - x(x+2)\left(\frac{1}{x+2}\right) = x(x+2)\frac{2}{x(x+2)}$

$\_\_\_ - \_\_\_ = \_\_\_$

☐ c. Finish solving for $x$. $\_\_\_ = \_\_\_$

$x = \_\_\_$

☐ d. Substitute the solution for $x$ in the original equation. $\_\_\_ = \_\_\_$

☐ e. Is the solution extraneous? ___

3. Use cross multiplication to solve the proportion $\frac{7}{5x} = \frac{15}{10}$ for $x$.

☑ a. Multiply the numerator of $\frac{\mathbf{7}}{5x}$ by the denominator of $\frac{15}{\mathbf{10}}$. $\frac{\mathbf{7}}{5x} \searrow \frac{15}{\mathbf{10}}$

$7 \cdot 10$

☐ b. Multiply the denominator of $\frac{7}{\mathbf{5x}}$ by the numerator of $\frac{\mathbf{15}}{10}$. $\frac{7}{\mathbf{5x}} \nearrow \frac{\mathbf{15}}{10}$

$\_\_\_ \cdot \_\_\_$

☐ c. Set the products equal to each other. $\_\_\_ = 70$

☐ d. Finish solving for $x$. $x = \_\_\_$

### *Answers to Sample Problems*

*Problem 1*

*a.* 12, 12

*b.* 12, 12, 12, 12

$3(x-2)$, 36

*c.* $3x - 6 + 36 = 10 - 2x$

$5x = -20$

$x = -4$

*d.* $\frac{-4-2}{4} + 3 = \frac{5}{6} - \frac{-4}{6}$

*e.* No

*Problem 2*

*b.* $2(x+2)$, $x$, 2

*c.* $2x + 4 - x$, 2

$x = -2$

*d.* $\frac{2}{-2} - \frac{1}{-2+2} = \frac{2}{-2(-2+2)}$

*e.* Yes

*Problem 3*

*b.* $5x$, 15 (in either order)

*c.* $75x$

*d.* $\frac{70}{75}$ or $\frac{14}{15}$

# HOMEWORK

## Homework Problems

Circle the homework problems assigned to you by the computer, then complete them below.

### Explain
### Solving Equations with Rational Expressions

In problems 1 through 12 solve for the variable. Be sure to say if the solution is extraneous.

1. Solve for $x$: $\frac{5}{x} - \frac{2}{x} = 1$
2. Solve for $y$: $\frac{4}{7}y = -\frac{2}{7}$
3. Solve for $x$: $\frac{3}{x} + \frac{2}{x-2} = 1$
4. Solve for $y$: $\frac{y}{y-5} - \frac{2}{5} = \frac{5}{y-5}$
5. Solve for $x$: $\frac{3x+1}{11x-9} = \frac{1}{3}$
6. Solve for $x$: $\frac{2}{1-x} - \frac{1}{x} = \frac{7}{6}$
7. Solve for $y$: $\frac{4y-9}{8} = \frac{6-8y}{16}$
8. Solve for $x$: $\frac{3}{2x-3} - x = \frac{2}{4x-6}$
9. A person who weighs 100 pounds on Earth would weigh 38 pounds on Mars. Use the proportion below to figure out how much someone who weighs 160 pounds on Earth would weigh on Mars.

$$\frac{\text{weight on Mars}}{\text{weight on Earth}} = \frac{38}{100} = \frac{x}{160}$$

10. A person who weighs 100 pounds on Earth would weigh 234 pounds on Jupiter. Use the proportion below to figure out how much someone who weighs 160 pounds on Earth would weigh on Jupiter.

$$\frac{\text{weight on Jupiter}}{\text{weight on Earth}} = \frac{234}{100} = \frac{x}{160}$$

11. Solve for $y$: $\frac{6}{x-2} - \frac{3}{x} = \frac{-5}{x-4}$
12. Solve for $x$: $\frac{6}{x-2} - \frac{1}{3} = \frac{3x}{x-2}$

# APPLY

## Practice Problems

Here are some additional practice problems for you to try.

### Solving Equations with Rational Expressions

1. Solve for $x$: $\frac{4}{x} + \frac{3}{x} = 1$
2. Solve for $x$: $\frac{8}{x} - \frac{4}{x} = 1$
3. Solve for $x$: $\frac{1}{x+1} + \frac{5}{x+1} = 2$
4. Solve for $x$: $\frac{10}{x-3} - \frac{4}{x-3} = -3$
5. Solve for $x$: $\frac{3}{x+2} + \frac{4}{x+2} = -1$
6. Solve for $x$: $\frac{x}{1-x} + \frac{3}{1-x} = -5$
7. Solve for $x$: $\frac{7}{x+3} - \frac{x}{x+3} = -3$
8. Solve for $x$: $\frac{4}{x-2} - \frac{x}{x-2} = -3$
9. Solve for $x$: $\frac{x-1}{4} + \frac{x}{3} = -2$
10. Solve for $x$: $\frac{x+2}{6} - \frac{x}{2} = 5$
11. Solve for $x$: $\frac{x+2}{3} + \frac{3x}{4} = 5$
12. Solve for $x$: $\frac{x-1}{5} + \frac{x}{3} = \frac{3x+2}{15}$
13. Solve for $x$: $\frac{x+3}{6} - \frac{x}{5} = \frac{9x-5}{30}$
14. Solve for $x$: $\frac{x+3}{5} + \frac{x}{4} = \frac{8x+4}{20}$
15. Solve for $x$: $\frac{4}{x} + \frac{2}{x+1} = 5$
16. Solve for $x$: $\frac{3}{x-2} - \frac{1}{x} = -4$
17. Solve for $x$: $\frac{2}{x} - \frac{1}{x-3} = 2$
18. Solve for $x$: $\frac{x}{5} + \frac{15}{3x} = \frac{x+3}{4}$
19. Solve for $x$: $\frac{24}{8x} - \frac{x}{3} = \frac{3-x}{6}$
20. Solve for $x$: $\frac{x}{6} + \frac{12}{2x} = \frac{x-2}{4}$
21. Solve for $x$: $\frac{x+1}{5} - \frac{3}{x} = \frac{x-2}{x}$
22. Solve for $x$: $\frac{3}{x} - \frac{x-2}{3} = \frac{x+1}{2x}$
23. Solve for $x$: $\frac{x-2}{7} - \frac{5}{x} = \frac{x+5}{x}$
24. Solve for $x$: $\frac{2}{3(x+4)} - \frac{4}{3} = \frac{2}{x}$
25. Solve for $x$: $\frac{6}{4(x-2)} + \frac{1}{4} = \frac{4}{x}$
26. Solve for $x$: $\frac{8}{5(x-3)} + \frac{1}{5} = \frac{5}{x}$
27. Solve for $x$: $5 - \frac{3}{x+4} = \frac{5x+20}{x+4}$
28. Solve for $x$: $\frac{2}{x-2} + 3 = \frac{3x-6}{x-2}$

# EVALUATE

## Practice Test

Take this practice test to be sure that you are prepared for the final quiz in Evaluate.

1. Solve $\frac{2}{3x} - \frac{1}{x} = \frac{1}{15}$ for $x$. Is the solution extraneous?

2. Solve $\frac{4y}{y+3} - \frac{1}{2} = \frac{9}{y+3}$ for $y$. Is the solution extraneous?

3. The volume $V$ of a right circular cone is $V = \frac{1}{3}\pi r^2 h$, where $r$ is the radius, and $h$ is the height. Solve this formula for $h$.

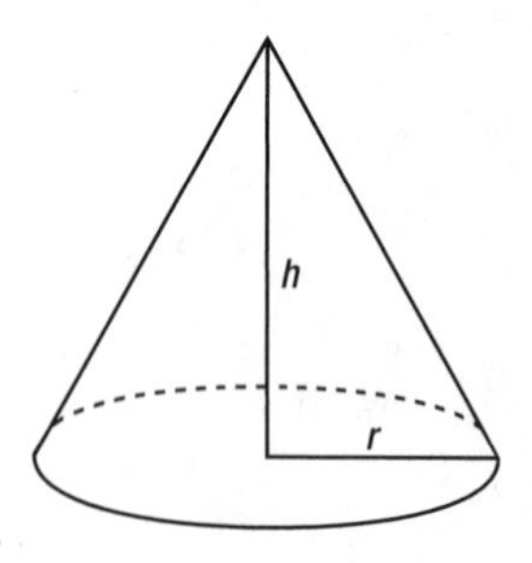


4. Solve this proportion for $x$: $\frac{5x-8}{2x+1} = \frac{4}{3}$

5. Solve $\frac{5}{4y} - \frac{2}{2y} = \frac{1}{16}$ for $y$. Is the solution extraneous?

6. Solve $\frac{x}{x+5} - \frac{1}{5} = \frac{3}{x+5}$ for $x$. Is the solution extraneous?

7. The surface area, $S$, of a right circular cylinder is $S = 2\pi rh + 2\pi r^2$, where $r$ is the radius, and $h$ is the height. Solve this formula for $h$.

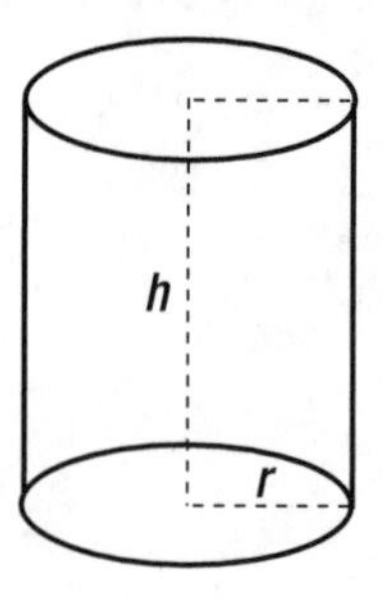


8. Solve this proportion for $y$: $\frac{6y-4}{6y+6} = \frac{4}{9}$

# ANSWERS

## Homework

**1.** $x = 3$ **3.** $x = 6$ or $x = 1$ **5.** $x = 6$ **7.** $y = \frac{3}{2}$

**9.** 60.8 lbs. **11.** $x = 3$ or $x = -1$

**12.** $x = 2$, the solution is extraneous

## Practice Problems

**1.** $x = 7$ **3.** $x = 2$ **5.** $x = -9$ **7.** $x = -8$ **9.** $x = -3$

**11.** $x = 4$ **13.** $x = 2$ **15.** $x = 1$ or $x = -\frac{4}{5}$

**17.** $x = 2$ or $x = \frac{3}{2}$ **19.** $x = 3$ or $x = -6$

**21.** $x = 5$ or $x = -1$ **23.** $x = 14$ or $x = -5$

**25.** $x = 4$ or $x = 8$ **27.** No solution

## Practice Test

**1.** $-5 = x$

The solution is not extraneous.

**2.** $y = 3$

The solution is not extraneous.

**3.** $\frac{3V}{\pi r^2} = h$

**4.** $x = 4$

**5.** $4 = y$

The solution is not extraneous.

**6.** $x = 5$

The solution is not extraneous.

**7.** $\frac{S}{2\pi r} - r = h$

**8.** $y = 2$

# LESSON 8.4 – PROBLEM SOLVING

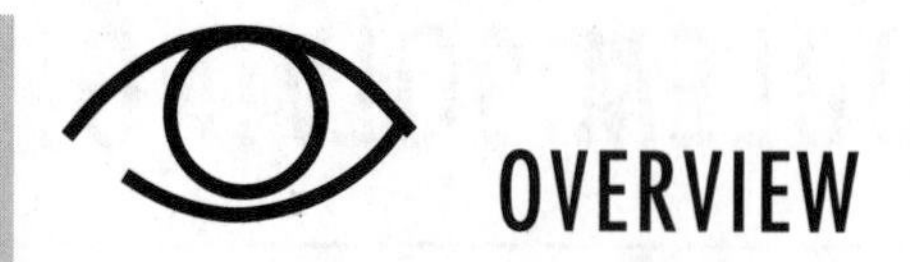

# OVERVIEW

***Here is what you'll learn in this lesson:***

***Using Rational Expressions***

*a. Ratio and proportion*

*b. Distance problems*

*c. Work problems*

*d. Variation*

Even a day at the beach can involve algebra. For example, have you ever wondered how much faster you could get across a cove by swimming instead of jogging along the shore? Or how bad a sunburn you'd get if you used SPF 6 instead of SPF 15 sunscreen? Or how much faster you could set up the volleyball court if two people worked together?

In this lesson, you'll apply what you've learned about rational expressions to solve a variety of problems like these.

# EXPLAIN

## RATIONAL EXPRESSIONS

### Summary

#### Solving Word Problems Involving Rational Expressions

In this lesson you will learn how to solve word problems when the equations contain rational expressions. Using equations with rational expressions allows you to solve a wide variety of problems.

To solve word problems using equations that contain rational expressions, you can often use this process:

1. Carefully read the problem and find the important information.
2. Determine what the problem is asking for and assign a variable to this unknown.
3. Set up an equation.
4. Solve the equation.
5. Check your answer in the original problem.

Below are some examples.

**Example 1** Working alone, it would take Michael 2 hours to paint a fence. If Beth worked alone, it would take her 3 hours to paint the fence. How long will it take them to paint the fence together?

1. Find the important information.

   Michael can paint the fence in 2 hours
   Beth can paint the fence in 3 hours

2. Assign the variable.

   Let $t$ = the amount of time it will take them to paint the fence together

3. Set up an equation.

   $$\text{fraction of the job Michael does} + \text{fraction of the job Beth does} = 1 \text{ complete job}$$

   $$\frac{t}{2} + \frac{t}{3} = 1$$

4. Solve the equation.

   $$6\left(\frac{t}{2} + \frac{t}{3}\right) = 6(1)$$

   $$3t + 2t = 6$$

   $$5t = 6$$

   $$t = \frac{6}{5}$$

*Where did the equation come from?*

*Well, Beth can paint the entire fence in 3 hours, so*

*in 1 hour she can paint $\frac{1}{3}$ of the fence.*

*in 2 hours she can paint $\frac{2}{3}$ of the fence.*

*in t hours she can paint $\frac{t}{3}$ of the fence.*

*Similarly, Michael can paint the entire fence in 2 hours, so*

*in 1 hour he can paint $\frac{1}{2}$ of the fence.*

*in t hours he can paint $\frac{t}{2}$ of the fence.*

5. Check your answer in the original problem.

Is $\text{fraction Michael paints} + \text{fraction Beth paints} = 1$?

Is $\dfrac{\frac{6}{5}}{2} + \dfrac{\frac{6}{5}}{3} = 1$?

Is $\frac{6}{5} \cdot \frac{1}{2} + \frac{6}{5} \cdot \frac{1}{3} = 1$?

Is $\frac{3}{5} + \frac{2}{5} = 1$?

Is $1 = 1$? Yes.

So, working together it would take them $\frac{6}{5}$ of an hour, or 1 hour and 12 minutes, to paint the fence.

**Example 2** Gretchen can run as fast as Petra. Last Saturday, Gretchen ran 4 miles farther than Petra ran. If Gretchen ran for 3 hours and Petra ran for 2 hours, how far did Petra go?

1. Find the important information.
   Gretchen and Petra run the same speed
   Petra ran for 2 hours
   Gretchen ran for 3 hours
   Gretchen ran 4 miles farther than Petra

2. Assign the variable.
   Let $d$ = the distance Petra ran

*If you solve the equation $d = r \cdot t$ for $r$ you get $r = \frac{d}{t}$. You know you want to solve for $r$ since the rates are equal.*

3. Set up an equation.
   Petra's rate = Gretchen's rate

$$\frac{\text{distance Petra ran}}{\text{time Petra ran}} = \frac{\text{distance Gretchen ran}}{\text{time Gretchen ran}}$$

$$\frac{d}{2} = \frac{d+4}{3}$$

4. Solve the equation.

$$\frac{d}{2} = \frac{d+4}{3}$$

$$3d = 2(d+4)$$

$$3d = 2d + 8$$

$$d = 8$$

5. Check your answer in the original problem.

Is Petra's rate = Gretchen's rate ?

Is $\frac{\text{distance Petra ran}}{\text{time Petra ran}} = \frac{\text{distance Gretchen ran}}{\text{time Gretchen ran}}$ ?

Is $\frac{8}{2} = \frac{12}{3}$ ?

Is $4 = 4$ ? Yes.

So, Petra ran 8 miles.

# Sample Problems

1. It would take Ernesto 2 hours to rake the yard by himself. It would take David 5 hours to rake the yard by himself. How long will it take them working together?

   ☑ a. Find the important information. — Ernesto takes 2 hours to rake. David takes 5 hours to rake

   ☑ b. Assign the variable. — Let $t$ = total time to rake the yard together

   ☐ c. Set up an equation. — ____________________

   ☐ d. Solve the equation. — $t$ = ____

   ☐ e. Check your answer in the original problem.

   Is ____________ = _______?

   Is ____________ = _______?

   Is ____________ = _______? _____

2. In octane, the ratio of hydrogen atoms to carbon atoms is 9 to 4. If there are 846 atoms of hydrogen in 47 molecules of octane, how many atoms of carbon are in these 47 molecules of octane?

   ☑ a. Find the important information. — The ratio of hydrogen atoms to carbon atoms is 9 to 4. 47 molecules of octane contain 846 hydrogen atoms.

   ☑ b. Assign the variable. — Let $x$ = the number of carbon atoms in 47 molecules of octane

   ☑ c. Set up an equation. — $\frac{\text{hydrogen atoms}}{\text{carbon atoms}} = \frac{9}{4} = \frac{846}{x}$

   ☐ d. Solve the equation. — $x$ = _____

   ☐ e. Check your answer in the original problem.

   Is ____________ = _______?

   Is ____________ = _______?

   Is ____________ = _______? _____

***Answers to Sample Problems***

*c.* $\frac{t}{2} + \frac{t}{5} = 1$

*d. Below is one way to solve the equation.*

$$10\left(\frac{t}{2} + \frac{t}{5}\right) = 10(1)$$

$$5t + 2t = 10$$

$$7t = 10$$

$$t = \frac{10}{7}$$

*So it would take them $\frac{10}{7}$ hours, or about 1 hour and 26 minutes, to rake the yard working together.*

*e.* Is $\frac{\frac{10}{7}}{2} + \frac{\frac{10}{7}}{5} = 1$?

Is $\frac{10}{7} \cdot \frac{1}{2} + \frac{10}{7} \cdot \frac{1}{5} = 1$?

Is $\frac{5}{7} + \frac{2}{7} = 1$?

Is $1 = 1$? Yes.

*d. Below is one way to solve the equation.*

$$9 \cdot x = 4 \cdot 846$$

$$9x = 3384$$

$$x = 376$$

*So, there are 376 carbon atoms in 47 molecules of octane.*

*e.* Is $\frac{9}{4} = \frac{846}{376}$?

Is $\frac{9}{4} = \frac{\cancel{2}^{1} \cdot 3 \cdot 3 \cdot \cancel{47}^{1}}{\cancel{2}_{1} \cdot 2 \cdot 2 \cdot \cancel{47}_{1}}$?

Is $\frac{9}{4} = \frac{9}{4}$? Yes.

3. The area of a triangle varies jointly with its base and its height. If a triangle of area 50 inches$^2$ has a base of 25 inches and a height of 4 inches, what is the base of a triangle whose height is 8 inches and whose area is 40 inches$^2$?

☑ a. Find the important information.

area varies jointly with base and height

when the base is 25 inches and the height is 4 inches, the area is 50 inches$^2$

☑ b. Assign the variable.

Let $x$ = base of the triangle
Let $y$ = area of the triangle
Let $z$ = height of the triangle
Let $k$ = the constant of variation

☐ c. Set up an equation.

$y = kxz$

________________

☐ d. First find $k$, then substitute this value back into the equation $y = kxz$ to find the base of the triangle.

$k =$ _____

base = ______

☐ e. Check your answer in the original problem.

Is __________ = ______?

Is __________ = ______?

Is __________ = ______? _____

***Answers to Sample Problems***

*c.* $50 = k(4)(25)$

*d.* $50 = k(100)$

$\frac{50}{100} = k$

$k = \frac{1}{2}$

$40 = \frac{1}{2}(x)8$

$40 = 4x$

$10 = x$

*So, the base is 10 inches.*

*e.* Is $\frac{1}{2} \cdot 10 \cdot 8 = 40$?

Is $\frac{1}{2} \cdot 80 = 40$?

Is $40 = 40$? Yes.

# HOMEWORK

## Homework Problems

Circle the homework problems assigned to you by the computer, then complete them below.

### Explain
### Rational Expressions

1. To fill up their swimming pool, the Johnsons decided to use both their high volume hose and their neighbor's regular garden hose. If they had used only their hose, it would have taken them 12 hours to fill the pool, but using both hoses it took them only 7 hours. How long would it have taken them to fill the pool using only their neighbor's hose?

2. On her camping trip, Li spent as much time hiking as she did rafting. She traveled 2.5 miles per hour when she was rafting and 3 miles per hour when she was hiking. If she went 3 miles more hiking than she did rafting, how far did she hike?

3. The ratio of jellybeans to gummy bears in a bag of candy is 7 to 2. If there are 459 pieces of candy in the bag, how many jellybeans are there?

4. The mass of an object varies jointly with its density and its volume. If 156 grams of iron has a volume of 20 $cm^3$ and a density of 7.8 $\frac{\text{grams}}{\text{cm}^3}$, what is the volume of 312 grams of iron with the same density?

5. To empty their swimming pool, the Johnsons decided to use both the regular drain and a pump. If it takes 15 hours for the pool to empty using the drain alone and 7 hours for the pool to empty using the pump alone, how long will it take for the pool to empty using both the drain and the pump?

6. One cyclist can ride 2 miles per hour faster than another cyclist. If it takes the first cyclist 2 hours and 20 minutes to ride as far as the second cyclist rides in 2 hours, how fast can each go?

7. The ratio of the length of a rectangle to its width is 3 to 1. If the perimeter of the rectangle is 32 inches, what are its dimensions?

8. The amount of energy that can be derived from particles varies directly with their mass. If $8.184 \cdot 10^{-14}$ Nm of energy is obtained from a particle whose mass is $9.1066 \cdot 10^{-31}$ kg, how much energy can be derived from particles whose mass is 0.001 kg?

9. Melanie and Alex have to prune all of the trees in their yard. Working alone, it would take Melanie 7 hours to do all of the pruning. It would take Alex 11 hours to do all of the pruning by himself. How long will it take them working together?

10. A bicyclist and a horseback rider are going the same speed. The rider stops after 11.1 miles. The bicyclist goes for another hour and travels a total of 18.5 miles. How fast is each one going?

11. Fish and game wardens can estimate the population of fish in a lake if they take a sample of fish, tag them, return them to the lake, take another sample of fish, and look at the ratio of tagged fish to untagged fish. If a warden tags 117 fish in the first sample, and then finds 13 out of 642 fish have tags in the second sample, how many fish were in the lake?

12. The height of a pyramid of constant volume is inversely proportional to the area of its base. If a pyramid of volume 300 $\text{meters}^3$ has a base area of 90 $\text{meters}^2$ and a height of 10 meters, what is the height of a pyramid whose base area is 100 $\text{meters}^2$?

# APPLY

## Practice Problems

Here are some additional practice problems for you to try.

### Rational Expressions

1. Working alone it would take Josie 4 hours to paint a room. It would take Curtis 5 hours to paint the same room by himself. How long would it take them to paint the room if they work together?

2. Before the library is remodeled, all of the books must be packed in boxes. Working alone, it would take Gail 15 workdays to do the packing. It would take Rob 18 workdays. How long will it take them working together?

3. Two computers are available to process a batch of data. The faster computer can process the batch in 36 minutes. If both computers run at the same time, they can process the batch in 20 minutes. How long would it take the slower computer to process the batch alone?

4. Two copy machines are available to print final exams. The faster copy machine can do the whole job in 75 minutes. If both machines print at the same time, they can do the whole job in 50 minutes. How long would it take the slower machine to do the whole job alone?

5. Two tomato harvesters are available to harvest a field of tomatoes. The slower harvester can harvest the whole field in 7 hours. If both machines harvest at the same time, they can harvest the whole field in 3 hours. How long would it take the faster machine to harvest the whole field by itself?

6. There are two overflow pipes at a dam. The larger overflow pipe can lower the level of the water in the reservoir by 1 foot in 45 minutes. The smaller one lowers the level of water by 1 foot in 2 hours 15 minutes. If both overflow pipes are open at the same time, how long will it take them to lower the level of water by 1 foot?

7. Two fire hoses are being used to flood the skating rink at the park. The larger hose alone can flood the park in 50 minutes. The smaller hose alone can flood the park in 1 hour and 15 minutes. If both hoses run at the same time, how long will it take them to flood the park?

8. Used by itself, the cold water faucet can fill a bathtub in 12 minutes. It takes 15 minutes for the hot water faucet to fill the bathtub. If both faucets are on, how long will it take to fill the bathtub?

9. A box of chocolates contains caramel chocolates and nougat chocolates. The ratio of the number of caramels in the box to the number of nougats in the box is 4 to 3. There are a total of 42 chocolates in the box. How many caramel chocolates are in the box? How many nougat chocolates are in the box?

10. A fast food stand sells muffins and cookies. Last Monday, the ratio of the number of muffins sold to the number of cookies sold was 16 to 13. A total of 145 muffins and cookies were sold. How many muffins were sold? How many cookies were sold?

11. At a certain animal shelter, the ratio of puppies to adult dogs is 7 to 4. This week, there are a total of 55 dogs in the shelter. How many puppies are in the shelter this week? How many adult dogs are in the shelter this week?

12. In a certain cookie recipe, the ratio of cups of flour to cups of sugar is 3 to 1. If the recipe uses $2\frac{1}{4}$ cups of flour, how much sugar does it use?

13. In one multivitamin pill, the ratio of the number of units of Vitamin C to the number of units of Vitamin E is 40 to 13. If the pill contains 200 units of Vitamin C, how many units of vitamin E does it contain?

14. The ratio of the amount of caffeine, in milligrams, in a 12-ounce serving of coffee to the amount of caffeine, in milligrams, in a 12-ounce serving of cola is 25 to 9. If a 12-ounce serving of cola contains 72 milligrams of caffeine, how much caffeine does a 12-ounce serving of coffee contain?

15. Jayme can ride his bike as fast as Terry. Each day, Jayme rides his bike for one hour and 20 minutes. Each day, Terry rides his bike for two hours and rides 15 miles further than Jayme. How far does each ride?

16. Saskia runs as fast as Tanya. Each day, Tanya runs for 40 minutes. Each day, Saskia runs for one hour and runs 2 miles farther than Tanya. How far does each run?

17. Leroy rows a boat as fast as Sasha rows a boat. If Leroy rows for 30 minutes, he travels 1 mile farther than Sasha when she rows for 20 minutes. How far does each row?

18. Pietro and Maria spend the same amount of time driving to school. Pietro averages 50 miles per hour and Maria averages 30 miles per hour. Pietro drives 10 miles farther than Maria. How far does each drive to school?

19. Ranji and Paula spend the same amount of time driving to work. Ranji averages 60 miles per hour and Paula averages 40 miles per hour. Ranji drives 15 miles farther than Paula. How far does each drive to work?

20. A car averages 55 miles per hour and an airplane averages 75 miles per hour. If the airplane and the car travel for the same amount of time, the airplane travels 100 miles farther than the car. How far does each travel?

21. The accuracy of a car's speedometer varies directly with the actual speed of the car. A car's speedometer reads 24 miles per hour when the car is actually traveling at 32 miles per hour. When the speedometer reads 51 miles per hour, how fast is the car actually going?

22. The force needed to stretch a spring a certain distance varies directly with the distance. An 8 pound force stretches a spring 3.5 inches. How much force is needed to stretch the spring 12 inches?

23. A person's weight on the moon varies directly as the person's weight on Earth. A person weighing 144 pounds on Earth weighs only 24 pounds on the moon. How much does a person weigh on Earth who weighs 30 pounds on the moon?

24. The current, $i$, in an electrical circuit with constant voltage varies inversely as the resistance, $r$, of the circuit. The current in a circuit with constant voltage is 5 amperes when the resistance is 8 ohms. What is the current in the circuit if the resistance is increased to 10 ohms?

25. For storage boxes with the same volume, the area of the bottom of the box varies inversely with the height of the box. The area of the bottom of the box is 108 square inches when the height is 20 inches. What is the area when the height is 16 inches?

26. The time it takes a car to travel a fixed distance varies inversely with the rate at which it travels. It takes the car 4 hours to travel a fixed distance when it travels at a rate of 50 miles per hour. How fast does the car have to travel to cover the same distance in $2\frac{1}{2}$ hours?

27. The volume of a gas is directly proportional to the temperature of the gas and inversely proportional to the pressure exerted on the gas. Write a formula expressing this property. Use $V$ for volume, $T$ for temperature, and $P$ for pressure.

28. The resistance of an electric wire is directly proportional to the length of the wire and inversely proportional to the square of its diameter. Write a formula expressing this property using $R$ for resistance, $L$ for length, and $D$ for diameter.

# EVALUATE

## Practice Test

Take this practice test to be sure that you are prepared for the final quiz in Evaluate.

1. Caleb and Daria are going to wash windows. Working alone, it would take Daria 4 hours to wash the windows. It would take Caleb 3 hours to wash the windows by himself. How long will it take them to wash the windows working together?

2. Trisha ran to the park and then walked home. It took her $\frac{1}{2}$ hour to get to the park and 1 hour and 20 minutes to get home. If she runs 5 miles an hour faster than she walks, how far does she live from the park?

3. The ratio of raisins to peanuts in a bag of party mix is 5 to 6. If the bag contains 462 items, how many peanuts are there?

4. The area of a kite varies jointly with the lengths of its two diagonals. If a kite with area 30 inches$^2$ has one diagonal of length 10 inches and the other diagonal of length 6 inches, what is the area of a kite with diagonals of length 8 inches and 13 inches?

5. Marta is helping Ned wash dishes after a big party. If Ned could do all of the dishes by himself in 60 minutes and Marta could do all of the dishes by herself in 90 minutes, how long will it take them to do the dishes working together?

6. A harpy eagle can fly 35 kilometers per hour faster than a ruby topaz hummingbird. In the same amount of time, an eagle can fly 8.5 kilometers and a hummingbird can fly 5 kilometers. How fast can each bird fly?

7. The ratio of roses to carnations that a florist ordered was 3 to 4. If the florist received a total of 441 flowers, how many of those were roses?

8. The speed of a wave varies jointly with the wavelength and the frequency of the wave. If the speed of a wave is 20 feet per second, its wavelength is 50 feet and its frequency is 0.4 waves per second. What is the speed of a wave whose wavelength is 1 foot and whose frequency is 8 waves per second?

# ANSWERS

## Homework

**1.** 16.8 hr **3.** 357 jellybeans

**5.** $\frac{105}{22}$ hours (approximately 4.8 hours or 4 hours and 46 minutes)

**7.** 4 inches by 12 inches

**9.** $\frac{77}{18}$ hours (approximately 4.3 hours or 4 hours and 17 minutes)

**11.** 5,778 fish

## Practice Problems

**1.** $\frac{20}{9}$ hours or approximately 2 hours 13 minutes

**3.** 45 minutes **5.** $\frac{21}{4}$ hours or 5 hours and 15 minutes

**7.** 30 minutes **9.** 24 caramels, 18 nougats

**11.** 35 puppies, 20 adult dogs **13.** 65 units

**15.** Jayme: 30 miles; Terry: 45 miles

**17.** Sasha: 2 miles; Leroy: 3 miles

**19.** Ranji: 45 miles, Paula: 30 miles

**21.** 68 miles per hour **23.** 180 pounds

**25.** 135 square inches **27.** $V = \frac{kT}{P}$

## Practice Test

**1.** It will take them $\frac{12}{7}$ hours, or about 1 hour and 43 minutes, to wash the dishes working together.

**2.** Trish lives 4 miles from the park.

**3.** There are 252 peanuts in the bag of mix.

**4.** The area of the kite is 52 inches$^2$.

**5.** It will take them 36 minutes working together.

**6.** The hummingbird can fly at 50 km per hour and the eagle can fly at 85 km per hour.

**7.** The florist received 189 roses.

**8.** The speed of the wave is 8 feet per second.

# TOPIC 8 CUMULATIVE ACTIVITIES

## CUMULATIVE REVIEW PROBLEMS

These problems combine all of the material you have covered so far in this course. You may want to test your understanding of this material before you move on to the next topic. Or you may wish to do these problems to review for a test.

1. Solve $-13 \le 5x - 3 < 4$ for $x$.
2. Find: $a^2b^3c \cdot ab^2c^3$
3. Factor: $a^2 - b^2$
4. Circle the true statements.

   $\frac{1}{2} + \frac{1}{3} = \frac{2}{5}$

   $|19 + 4| = |19| + |4|$

   If $R = \{1, 2, 3\}$ and $S = \{1, 2, 3, 4, 5\}$, then $R \subset S$.

   $7 + 3 \cdot 6 = 60$

   $\frac{56}{63} = \frac{8}{9}$

   The GCF and LCM of two numbers is usually the same.
5. Use the Pythagorean Theorem to find the distance between the points $(2, -5)$ and $(-3, 7)$. See Figure 8.1.

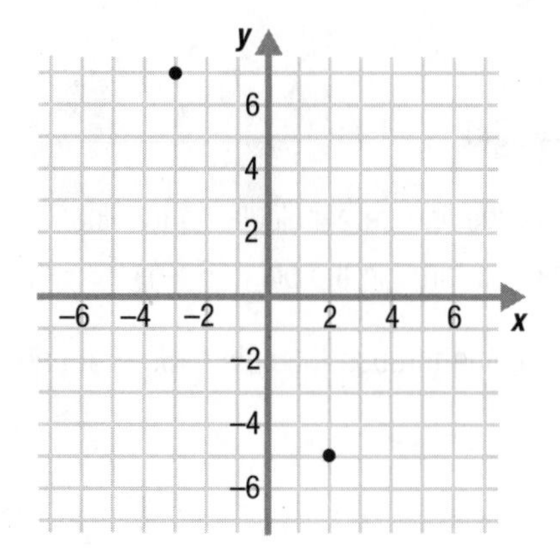


*Figure 8.1*

6. Factor: $18ab^4c^3 + 9a^4b^3c^2 + 12ab^2c^5$
7. Find the equation of the line parallel to the line $y = 3x + 2$ that passes through the point $(-2, 5)$.
8. Find: $(4a^2b + 3a - 9b) - (7a + 2b - 8a^2b)$
9. Solve $\frac{5}{x+3} - \frac{3}{x+3} = 1$ for $x$.
10. Write in scientific notation:

    a) 42,789,400

    b) 0.0025815
11. Find the slope and $y$-intercept of this line: $9x + 5y = 11$
12. Find:

    a. $3^4 \cdot 3$

    b. $\frac{a}{a^9}$

    c. $(x^7y^0)^5$
13. It would take Kendra 4 hours to type a report. It would take Gerri $2\frac{1}{2}$ hours to type the same report. How long would it take them to type the report working together?
14. Write the equation of the circle with radius 3 whose center is at $(1, 5)$.
15. Find: $\frac{3}{x-2} + \frac{1}{x-3}$
16. Factor: $6ab - 10a + 9b - 15$
17. Factor: $x^4 - y^4$
18. Evaluate the expression $5x^2 - 6xy^4 - 4 + 7y$ when $x = 3$ and $y = 0$.
19. Solve $2(3 + y) = 5(\frac{2}{5}y + 1)$ for $y$.
20. Solve for $x$: $\frac{2}{x} - \frac{1}{x-1} = \frac{5}{2x}$
21. Use the distance formula to find the square of the distance between the points $(1, 1)$ and $(7, -4)$.
22. For what values of $x$ is the expression $\frac{7}{x^2-9}$ undefined?

23. Find: $(9a^8b^3c - 12a^4b^3c^6) \div 3a^7b^3c^4$

Use Figure 8.2 to answer questions 24 through 26.

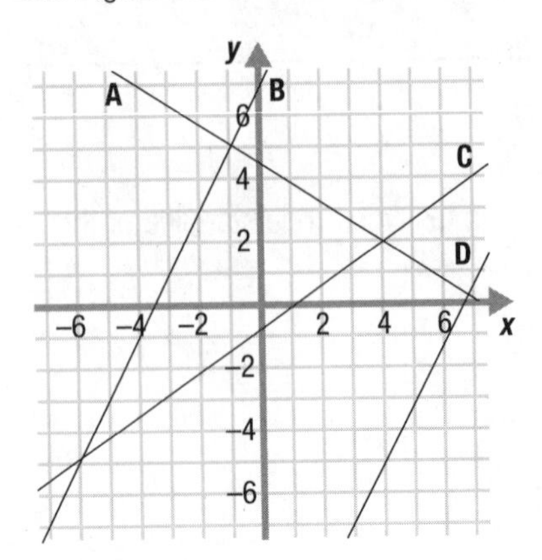


*Figure 8.2*

24. Which two lines form a system that has a solution of (4, 2)?

25. Which two lines form a system that has a solution of (–1, 5)?

26. Which two lines form a system that has no solution?

27. Graph the inequality $\frac{6}{5}x + 2y \geq 1$.

28. The ratio of dark brown candies to light brown candies in a bag is 3 to 1. If there are 53 light brown candies in the bag, how many are dark brown?

29. Find the slope of the line through the points (6, –2) and (4, –11).

30. Find: $-2s^3t(7r^2st^5 - r^3t)$

31. Factor: $6a^3b^2 - 24a^2b^3 + 24ab^2$

32. Graph the system of inequalities below to find its solution.

$$y \leq \frac{2}{3}x + 3$$

$$y > -3x - 4$$

33. Find: $\frac{x^2 - x - 6}{x^2 - 25} \cdot \frac{x^2 - 5x}{x^2 - x - 6}$

34. Factor: $8x^2y - 6xy^2 + 12x - 9y$

35. Solve $2 - \frac{x}{x+3} = \frac{-3}{x+3}$ for $x$.

36. Find the slope of the line that is perpendicular to the line that passes through the points (8, –9) and (–3, 11).

37. Find the equation of the line through the point (–2, 6) that has slope $\frac{4}{3}$:

    a. in point-slope form.

    b. in slope-intercept form.

    c. in standard form.

38. Emily withdrew \$985 in 5-dollar and 20-dollar bills from her savings account. If she had 65 bills altogether, how many of each did she have?

39. Find:

    a. $2x^0 - 3x^0 + 4x^0$

    b. $(x^0 \cdot x^0 \cdot x^0)^2$

    c. $\frac{a^2 \cdot c}{a^5 \cdot b^6 \cdot c^4}$

40. Factor: $2x^2 + xy - 3y^2$

41. Find the radius and the center of the circle whose equation is:

$$(x - 2)^2 + (y + 6)^2 = 25$$

42. Find: $\frac{23}{\frac{9}{4^3} + \frac{5}{6^2}}$

43. Factor: $49x^2 - 14x + 1$

44. Find: $(a^2b^2 - 4a^2b - 4ab^2 + 16ab + 2b - 8) \div (b - 4)$

45. Solve $3(x + 2) = 3x + 6$ for $x$.

46. Graph the line $x = -2.5$.

47. Find the equation of the line perpendicular to the line $y = 2x - 5$ that passes through the point (6, –1).

48. Find the slope of the line through the points (6, –14) and (22, 17).

49. Factor: $x^2 - 4x + 3$

50. Write in expanded form:

    a. $7.1047 \cdot 10^{12}$

    b. $4.294036 \cdot 10^{-8}$

# ANSWERS

## Cumulative Review Problems

**1.** $-2 \le x < \frac{7}{5}$ **3.** $(a+b)(a-b)$ **5.** 13 units

**7.** $y = 3x + 11$ **9.** $x = -1$

**11.** $m = -\frac{9}{5}$, $y$-intercept $= (0, \frac{11}{5})$

**13.** $t = \frac{20}{13}$ hours $\approx$ 1 hour and 32 minutes

**15.** $\frac{4x-11}{(x-2)(x-3)}$ or $\frac{4x-11}{x^2-5x+6}$ **17.** $(x^2+y^2)(x+y)(x-y)$

**19.** no solution **21.** 61 units

**23.** $\frac{3a}{c^3} - \frac{4c^2}{a^3}$ or $\frac{3a^4-4c^5}{a^3c^3}$ **25.** A and B

**27.**

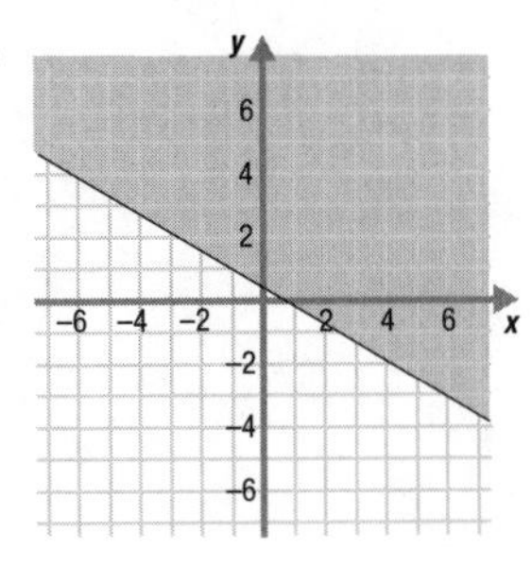


**29.** $m = \frac{9}{2}$ **31.** $6ab^2(a^2 - 4ab + 4)$ **33.** $\frac{x}{x+5}$

**35.** $x = -9$ **37a.** $y - 6 = \frac{4}{3}(x+2)$ **b.** $y = \frac{4}{3}x + \frac{26}{3}$

**c.** $-4x + 3y = 26$

**39a.** 3 **b.** 1 **c.** $\frac{1}{a^3b^6c^3}$ or $a^{-3}b^{-6}c^{-3}$

**41.** radius: 5; center: $(2, -6)$

**43.** $(7x-1)(7x-1)$ or $(7x-1)^2$ **45.** infinite solutions

**47.** $y = -\frac{1}{2}x + 2$ **49.** $(x-3)(x-1)$

# TOPIC 8 INDEX

# LESSON 9.1 – RATIONAL EXPONENTS

# OVERVIEW

*Here's what you'll learn in this lesson:*

***Roots and Exponents***

*a. The $n^{th}$ root of a number*

*b. Definition of $a^{\frac{1}{n}}$ and $a^{\frac{m}{n}}$*

*c. Properties of rational exponents*

***Simplifying Radicals***

*a. Simplifying radicals*

***Operations on Radicals***

*a. Adding and subtracting radical expressions*

*b. Multiplying radical expressions*

*c. Dividing radical expressions*

A farmer is experimenting with different fertilizers and varieties of corn in order to find ways to boost crop production. A researcher is studying the effects of pollutants and disease on fish populations around the world. A student volunteer is analyzing surveys to help increase donation levels for his organization.

Each of these people—Emerson Sarawop, the farmer; Sharon Ming, the researcher; and Vince Poloncic, the student volunteer—does work for the Center for World Hunger. And, each works every day with equations that involve radicals.

In this lesson, you will learn about radicals. You will learn how to simplify expressions that contain radicals. You will also learn how to add, subtract, multiply, and divide such expressions.

# EXPLAIN

## ROOTS AND EXPONENTS

### Summary

#### Square Roots

When you square the square root of a number, you get back the original number.

For example:

$$\left(\sqrt{16}\right)^2 = 16$$

*The square root of 16 is written as $\sqrt{16}$.*

Every positive integer has both a positive and a negative square root. The symbol $\sqrt{a}$ denotes the positive square root of *a*. The symbol $-\sqrt{a}$ denotes the negative square root of *a*.

For example:

$\sqrt{25} = 5$ $\quad$ $-\sqrt{25} = -5$

$\sqrt{16} = 4$ $\quad$ $-\sqrt{16} = -4$

$\sqrt{81} = 9$ $\quad$ $-\sqrt{81} = -9$

*The positive square root is called the principal square root.*

You can't take the square root of a negative number and get a real number because no real number times itself equals a negative number.

#### Cube Roots

When you cube the cube root of a number, you get back the original number.

For example:

$$\left(\sqrt[3]{64}\right)^3 = 64$$

*The cube root of 64 is written as $\sqrt[3]{64}$.*

Both positive and negative numbers have real cube roots.

For example:

$\sqrt[3]{125} = 5$ $\quad$ because $5 \cdot 5 \cdot 5 = 125$

$\sqrt[3]{-125} = -5$ $\quad$ because $(-5) \cdot (-5) \cdot (-5) = -125$

## $n$th Roots

Numbers also have 4th roots, 5th roots, 6th roots, and so on.

For example:

$\sqrt[4]{81} = 3$ because $3 \cdot 3 \cdot 3 \cdot 3 = 81$

$\sqrt[5]{-1} = -1$ because $(-1) \cdot (-1) \cdot (-1) \cdot (-1) \cdot (-1) = -1$

$\sqrt[6]{64} = 2$ because $2 \cdot 2 \cdot 2 \cdot 2 \cdot 2 \cdot 2 = 64$

When you raise the $n$th root of a number to the $n$th power, you get back the original number.

For example:

$$\left(\sqrt[n]{10}\right)^n = 10$$

In general, the $n$th root of a number $a$ is written

$$\sqrt[n]{a}$$

where $n$ is a positive integer.

index → $\sqrt[n]{a}$ ← radicand

Here, $a$ is called the radicand and $n$ is called the index. The index is the number of times that the root has to be multiplied in order to get the radicand.

*When there isn't any number written as the index, it is understood to be 2. So $\sqrt{a}$ is the same as $\sqrt[2]{a}$.*

When finding real roots:

- If $n$ is odd, then $\sqrt[n]{a}$ is a real number.
- If $n$ is even, then $\sqrt[n]{a}$ is a real number if $a \geq 0$.

For example:

$\sqrt[3]{8} = 2$ because $2^3 = 2 \cdot 2 \cdot 2 = 8$

$\sqrt[3]{-8} = -2$ because $(-2)^3 = (-2) \cdot (-2) \cdot (-2) = -8$

$\sqrt{36} = 6$ because $6^2 = 6 \cdot 6 = 36$

$\sqrt{-36}$ is not a real number

*If the index is even, then the radicand must be positive in order to get a real number. This is because there is no real number that, when multiplied by itself an even number of times, gives a negative number.*

## Rational Exponents

All roots can also be written as rational, or fractional, exponents.

In general:

$$\sqrt[n]{a} = a^{\frac{1}{n}}$$

For example:

$$\sqrt[3]{8} = 8^{\frac{1}{3}}$$

*You may find it easier to solve problems if you first rewrite the exponent with a radical sign. For example,*

$$16^{\frac{1}{2}} = \sqrt{16} = 4$$

Since you can rewrite rational exponents as roots, the same rules that apply to roots also apply to rational exponents:

- If $n$ is odd, then $a^{\frac{1}{n}}$ is a real number.
- If $n$ is even, then $a^{\frac{1}{n}}$ is a real number when $a \geq 0$.

If the numerator of the rational exponent is not equal to 1, you can still rewrite the problem using radicals.

In general:

$$a^{\frac{m}{n}} = \left(\sqrt[n]{a}\right)^m = \sqrt[n]{a^m}$$

*Notice that you get the same answer whether you first take the root of the number and then raise it to the appropriate power, or whether you first raise the radicand to the appropriate power and then take the root.*

To simplify an expression when the rational exponent is not equal to 1:

1. Rewrite the problem using radicals.
2. Take the appropriate root.
3. Raise the result to the correct power.

For example, to find $32^{\frac{2}{5}}$:

1. Rewrite the problem using radicals. $= \left(\sqrt[5]{32}\right)^2$
2. Take the 5th root. $= 2^2$
3. Simplify. $= 4$

*When dealing with large numbers, you may find it easier to first take the root of the number and then raise it to the correct power.*

Always reduce a rational exponent to lowest terms or you may get the wrong answer.

For example:

$(-16)^{\frac{1}{2}} = \sqrt{-16}$, which is not a real number

$(-16)^{\frac{2}{4}} \neq \sqrt[4]{(-16)^2} \neq \sqrt[4]{256} \neq 4$

Since $\frac{2}{4}$ is not reduced to lowest terms, the answer, 4, is incorrect.

The basic properties for integer exponents also hold for rational exponents as long as the expressions represent real numbers.

| Property of Exponents | Integer Exponents | Rational Exponents |
|---|---|---|
| Multiplication | $7^3 \cdot 7^5 = 7^{3+5} = 7^8$ | $7^{\frac{1}{2}} \cdot 7^{\frac{1}{4}} = 7^{\frac{1}{2}+\frac{1}{4}} = 7^{\frac{3}{4}}$ |
| Division | $\frac{3^2}{3^6} = 3^{2-6} = 3^{-4}$ | $\frac{3^{\frac{1}{2}}}{3^{\frac{1}{4}}} = 3^{\frac{1}{2}-\frac{1}{4}} = 3^{\frac{1}{4}}$ |
| Power of a Power | $(2^3)^4 = 2^{3 \cdot 4} = 2^{12}$ | $(5^{\frac{1}{2}})^{\frac{1}{3}} = 5^{\frac{1}{2} \cdot \frac{1}{3}} = 5^{\frac{1}{6}}$ |
| Power of a Product | $(5 \cdot 7)^2 = 5^2 \cdot 7^2$ | $(5 \cdot 7)^{\frac{2}{9}} = 5^{\frac{2}{9}} \cdot 7^{\frac{2}{9}}$ |
| Power of a Quotient | $\left(\frac{3}{8}\right)^4 = \frac{3^4}{8^4}$ | $\left(\frac{3}{8}\right)^{\frac{1}{4}} = \frac{3^{\frac{1}{4}}}{8^{\frac{1}{4}}}$ |

The properties of exponents can help you simplify some expressions.

For example, to simplify $(8 \cdot 27)^{\frac{1}{3}}$:

1. Apply the power of a product property. $= (8)^{\frac{1}{3}} \cdot (27)^{\frac{1}{3}}$
2. Rewrite the problem using radicals. $= \sqrt[3]{8} \cdot \sqrt[3]{27}$
3. Take the cube roots. $= 2 \cdot 3$
4. Simplify. $= 6$

*Answers to Sample Problems*

*a. −11*

*c. $\frac{\sqrt{81}}{\phantom{0}}$*

*d. $\frac{9}{10}$*

*d. 10*

## Sample Problems

1. Find: $\sqrt[3]{-1331}$
   - ☐ a. Simplify $\sqrt[3]{-1331}$. ________
2. Rewrite as a radical and evaluate: $\left(\frac{81}{100}\right)^{\frac{2}{4}}$
   - ☑ a. Reduce the exponent to lowest terms. $= \left(\frac{81}{100}\right)^{\frac{1}{2}}$
   - ☑ b. Apply the power of a quotient property of exponents. $= \frac{81^{\frac{1}{2}}}{100^{\frac{1}{2}}}$
   - ☐ c. Rewrite as radicals. = ________
   - ☐ d. Take the square root of the numerator and the denominator. = ________
3. Evaluate: $(8 \cdot 125)^{\frac{1}{3}}$
   - ☑ a. Raise each term to the $\frac{1}{3}$ power. $8^{\frac{1}{3}} \cdot 125^{\frac{1}{3}}$
   - ☑ b. Express exponents as radicals. $= \sqrt[3]{8} \cdot \sqrt[3]{125}$
   - ☑ c. Simplify the radicals. $= 2 \cdot 5$
   - ☐ d. Simplify. = ________

# SIMPLIFYING RADICALS

## Summary

Equations often contain radical expressions. In order to simplify these expressions, you have to know how to simplify radicals.

### The Multiplication Property of Radicals

The rule for multiplying square roots is:

The square root of a product = the product of the square roots.

For example:

$$\sqrt{144 \cdot 121}$$
$$= \sqrt{144} \cdot \sqrt{121}$$
$$= 12 \cdot 11$$
$$= 132$$

In general:

$$\sqrt[n]{a \cdot b} = \sqrt[n]{a} \cdot \sqrt[n]{b}$$

Here, $a$ and $b$ are real numbers, $\sqrt[n]{a}$ and $\sqrt[n]{b}$ are real numbers, and $n$ is a positive integer.

### Division Property of Radicals

The rule for dividing square roots is:

The square root of a quotient = the quotient of the square roots.

For example:

$$\sqrt{\frac{16}{169}}$$
$$= \frac{\sqrt{16}}{\sqrt{169}}$$
$$= \frac{4}{13}$$

In general:

$$\sqrt[n]{\frac{a}{b}} = \frac{\sqrt[n]{a}}{\sqrt[n]{b}}$$

Here, $a$ and $b$ are real numbers, $\sqrt[n]{a}$ and $\sqrt[n]{b}$ are real numbers, and $n$ is a positive integer.

## Sums and Differences of Roots

The $n$th root of a sum is not equal to the sum of the $n$th roots.

For example:

Is $\sqrt{9+16} = \sqrt{9} + \sqrt{16}$ ?

Is $\sqrt{25} = 3 + 4$ ?

Is $5 = 7$ ? No.

Similarly, the $n$th root of a difference is not equal to the difference of the $n$th roots.

## The Relationship Between Powers and Roots

If you start with a number, cube it, then take its cube root, you end up with the same number that you started with.

For example,

$$\sqrt[3]{8^3} = 8$$

However, if you start with a number, square it, then take its square root, you only get back the original number if the original number is greater than or equal to 0. If the original number is less than 0, taking the root will give you (–1) times the number.

For example:

$$\sqrt{9^2} = 9$$

$$\sqrt{(-9)^2} = 9 = (-1) \cdot (-9)$$

When taking roots:

- If the radicand is positive, $\sqrt[n]{a^n} = a$
- If $a$ is negative and $n$ is odd, $\sqrt[n]{a^n} = a$
- If $a$ is negative and $n$ is even, $\sqrt[n]{a^n} = -a$

## Simplifying Radicals

A radical expression is in simplest terms if it meets the following conditions:

- In the expression $\sqrt[n]{a^n}$, $a$ contains no factors (other than 1) that are raised to a power of $n$.
- There are no fractions under the radical sign.
- There are no radicals in the denominator of the expression.

To simplify a radical expression that contains factors which are powers of the index, $n$:

1. Write the radicand as a product of its prime factors.
2. Rewrite the factors using exponents.
3. Where possible, rewrite factors as a product having the index, $n$, as an exponent.
4. Bring all possible factors outside the radical.
5. Simplify.

For example, to simplify $\sqrt[3]{80}$:

1. Write 80 as a product of its prime factors. $= \sqrt[3]{2 \cdot 2 \cdot 2 \cdot 2 \cdot 5}$
2. Rewrite the factors using exponents. $= \sqrt[3]{2^4 \cdot 5}$
3. Rewrite $2^4$ as a product including $2^3$. $= \sqrt[3]{2^3 \cdot 2 \cdot 5}$
4. Bring $\sqrt[3]{2^3}$ outside the radical. $= 2\sqrt[3]{2 \cdot 5}$
5. Simplify. $= 2\sqrt[3]{10}$

To simplify a radical expression that has a fraction under the radical sign:

1. Rewrite the fraction with two radical signs—one in the numerator and one in the denominator.
2. Multiply the numerator and denominator of the fraction by the same number to eliminate the radical in the denominator of the fraction.
3. Simplify.

*When you multiply the numerator and denominator of a fraction by the same number, it is the same as multiplying the expression by 1, so the value of the rational expression doesn't change.*

For example, to simplify $\sqrt{\frac{2}{3}}$:

1. Rewrite the fraction with two radical signs. $= \frac{\sqrt{2}}{\sqrt{3}}$
2. Multiply the numerator and denominator by $\sqrt{3}$. $= \frac{\sqrt{2}}{\sqrt{3}} \cdot \frac{\sqrt{3}}{\sqrt{3}}$
3. Simplify. $= \frac{\sqrt{2} \cdot \sqrt{3}}{\sqrt{3} \cdot \sqrt{3}}$

$$= \frac{\sqrt{6}}{3}$$

To simplify a radical expression that has a radical in the denominator:

1. Multiply the numerator and denominator of the fraction by the same number to eliminate the radical in the denominator of the fraction.
2. Simplify.

*Why do you multiply $\sqrt[3]{5}$ by $\sqrt[3]{5^2}$? Because this gives $\sqrt[3]{5^3}$, which equals 5.*

For example, to simplify $\frac{7}{\sqrt[3]{5}}$:

1. Multiply the numerator and denominator by $\sqrt[3]{5^2}$. $= \frac{7}{\sqrt[3]{5}} \cdot \frac{\sqrt[3]{5^2}}{\sqrt[3]{5^2}}$

2. Simplify. $= \frac{7\sqrt[3]{5^2}}{\sqrt[3]{5 \cdot 5^2}}$

$= \frac{7\sqrt[3]{5^2}}{\sqrt[3]{5^3}}$

$= \frac{7\sqrt[3]{25}}{5}$

When simplifying radicals, it is helpful to recognize some perfect squares and perfect cubes. You may want to remember the numbers in this table:

| Number ($n$) | Square ($n^2$) | Cube ($n^3$) |
|---|---|---|
| 1 | 1 | 1 |
| 2 | 4 | 8 |
| 3 | 9 | 27 |
| 4 | 16 | 64 |
| 5 | 25 | 125 |
| 6 | 36 | 216 |
| 7 | 49 | 343 |
| 8 | 64 | 512 |
| 9 | 81 | 729 |
| 10 | 100 | 1000 |

***Answers to Sample Problems***

*b. $\frac{7}{8}$*

*c. $-5x^2y\sqrt[3]{y^2}$*

## Sample Problems

1. Simplify: $\sqrt{\frac{49}{64}}$

☑ a. Rewrite the fraction using two radical signs. $= \frac{\sqrt{49}}{\sqrt{64}}$

☐ b. Simplify the square roots. = ________

2. Simplify: $\sqrt[3]{-125x^6y^5}$

☑ a. Rewrite the radicand as a product of its prime factors. $= \sqrt{(-5)(-5)(-5)x^6y^5}$

☑ b. Rewrite the factors using cubes, where possible. $= \sqrt[3]{(-5)^3(x^2)^3y^3y^2}$

☐ c. Bring all perfect cubes outside the radical. = ________

# OPERATIONS ON RADICALS

## Summary

### Identifying Like Radical Terms

To add or subtract radical expressions or to eliminate a radical sign in the denominator of a fraction, you will need to identify similar, or like, radical terms.
Similar, or like, radical terms have the same index and the same radicand.

For example, here are two terms that are like terms:

| | |
|---|---|
| $\sqrt[3]{7}$ | index: 3; radicand: 7 |
| $4\sqrt[3]{7}$ | index: 3; radicand: 7 |

Here are two terms that are not like terms:

| | |
|---|---|
| $2\sqrt[4]{5}$ | index: 4; radicand: 5 |
| $\sqrt{5}$ | index: 2; radicand: 5 |

Here are two more terms that are not like terms:

| | |
|---|---|
| $3\sqrt[5]{9}$ | index: 5; radicand: 9 |
| $6\sqrt[5]{8}$ | index: 5; radicand: 8 |

### Adding and Subtracting Radical Expressions

Now that you can identify like terms you can add and subtract radical expressions.

For example, to find $5\sqrt[3]{54} + 8\sqrt[3]{2} - 5\sqrt[3]{250}$:

1. Factor the radicands into their prime factors. $= 5\sqrt[3]{2 \cdot 3 \cdot 3 \cdot 3} + 8\sqrt[3]{2} - 5\sqrt[3]{2 \cdot 5 \cdot 5 \cdot 5}$
2. Rewrite the factors using cubes, where possible. $= 5\sqrt[3]{2 \cdot 3^3} + 8\sqrt[3]{2} - 5\sqrt[3]{2 \cdot 5^3}$
3. "Undo" the perfect cubes. $= \left(3 \cdot 5\sqrt[3]{2}\right) + 8\sqrt[3]{2} - \left(5 \cdot 5\sqrt[3]{2}\right)$
4. Simplify. $= 15\sqrt[3]{2} + 8\sqrt[3]{2} - 25\sqrt[3]{2}$
5. Combine like terms. $= -2\sqrt[3]{2}$

*All the radicals in step (3) are like terms because they have the same index, 3, and the same radicand, 2.*

### Multiplying Radical Expressions

You can use the multiplication property of radicals to simplify complex radical expressions.

For example, to simplify $3\sqrt[4]{8} \cdot 5\sqrt[4]{4}$:

1. Apply the multiplication property of radicals. $= 3 \cdot 5\sqrt[4]{8 \cdot 4}$

2. Factor the radicands into their prime factors. $= 3 \cdot 5\sqrt[4]{2 \cdot 2 \cdot 2 \cdot 2 \cdot 2}$

3. Write the 2's as a product involving a factor of $2^4$. $= 15\sqrt[4]{2^4 \cdot 2}$

4. Bring $\sqrt[4]{2^4}$ outside the radical. $= 15 \cdot 2\sqrt[4]{2}$

5. Simplify. $= 30\sqrt[4]{2}$

Sometimes when you multiply polynomials, you use the distributive property. This property is also useful when you multiply radicals.

For example, to simplify $\sqrt{8}\left(4\sqrt{2} + 3\sqrt{24}\right)$:

1. Apply the distributive property. $= \sqrt{8}(4\sqrt{2}) + \sqrt{8}\left(3\sqrt{24}\right)$

2. Apply the multiplication property of radicals. $= 4\sqrt{8 \cdot 2} + 3\sqrt{8 \cdot 24}$

3. Simplify.

$$= 4\sqrt{16} + 3\sqrt{192}$$

$$= 4\sqrt{4 \cdot 4} + 3\sqrt{2 \cdot 2 \cdot 2 \cdot 2 \cdot 2 \cdot 2 \cdot 3}$$

$$= 4\sqrt{4^2} + 3\sqrt{(2^3)^2 \cdot 3}$$

$$= 4\sqrt{4^2} + 3\sqrt{8^2 \cdot 3}$$

$$= (4 \cdot 4) + \left(3 \cdot 8\sqrt{3}\right)$$

$$= 16 + 24\sqrt{3}$$

*Remember FOIL?*

*$(x + 1)(x + 2)$*

*$= (x \cdot x) + (x \cdot 2) + (1 \cdot x) + (1 \cdot 2)$*

*$= x^2 + 2x + x + 2$*

*$= x^2 + 3x + 2$*

You can also use the FOIL method to multiply radicals.

For example, to find $(\sqrt{x} + 2)(3\sqrt{x} - 7)$:

1. Find the sum of the products of the first terms, the outer terms, the inner terms, and the last terms. $= \left(\sqrt{x} \cdot 3\sqrt{x}\right) - \left(\sqrt{x} \cdot 7\right) + \left(2 \cdot 3\sqrt{x}\right) - (2 \cdot 7)$

2. Simplify.

$$= 3x - 7\sqrt{x} + 6\sqrt{x} - 14$$

$$= 3x - \sqrt{x} - 14$$

## Conjugates

Sometimes when you multiply two irrational numbers you end up with a rational number.

For example, to find $\left(2 + \sqrt{7}\right)\left(2 - \sqrt{7}\right)$:

1. Find the sum of the products of the first terms, the outer terms, the inner terms, and the last terms. $= (2)^2 - 2\sqrt{7} + 2\sqrt{7} - \left(\sqrt{7}\right)^2$

2. Simplify. $= (2)^2 - \left(\sqrt{7}\right)^2$

$= 4 - 7$

$= -3$

The expressions $2 + \sqrt{7}$ and $2 - \sqrt{7}$ are called conjugates of each other. When conjugates are multiplied, the result is a rational number.

In general, these expressions are conjugates of one another:

$\left(a + \sqrt{b}\right)$ and $\left(a - \sqrt{b}\right)$

$\left(\sqrt{a} + \sqrt{b}\right)$ and $\left(\sqrt{a} - \sqrt{b}\right)$

Here, $\sqrt{a}$ and $\sqrt{b}$ are real numbers.

When you multiply conjugates, here's what happens:

$$\left(\sqrt{a} + \sqrt{b}\right)\left(\sqrt{a} - \sqrt{b}\right)$$

$$= \left(\sqrt{a} \cdot \sqrt{a}\right) - \left(\sqrt{a} \cdot \sqrt{b}\right) + \left(\sqrt{b} \cdot \sqrt{a}\right) - \left(\sqrt{b} \cdot \sqrt{b}\right)$$

$$= \left(\sqrt{a}\right)^2 - \left(\sqrt{b}\right)^2$$

$$= a - b$$

As another example, to find $\left(\sqrt{6} + \sqrt{14}\right)\left(\sqrt{6} - \sqrt{14}\right)$:

1. Find the sum of the products of the first, outer, inner, and last terms. $= \left(\sqrt{6}\right)^2 - \left(\sqrt{14}\right)^2$

2. Simplify. $= 6 - 14$

$= -8$

## Dividing Radical Expressions

You can use the division property of radicals to simplify radical expressions.

For example, to simplify $\sqrt{\frac{7}{2y}}$:

1. Apply the division property of radicals. $= \frac{\sqrt{7}}{\sqrt{2y}}$

2. Rationalize the denominator. $= \frac{\sqrt{7}}{\sqrt{2y}} \cdot \frac{\sqrt{2y}}{\sqrt{2y}}$

3. Perform the multiplication. $= \frac{\sqrt{7 \cdot 2y}}{2y}$

4. Simplify. $= \frac{\sqrt{14y}}{2y}$

*Since the expression in the denominator is a square root, to eliminate it you must multiply it by itself one time (so there are a total of two factors of 2y under the square root sign). If the expression in the denominator had been $\sqrt[3]{2y}$, to eliminate it you would have had to multiply it by itself two times (so there would be a total of three factors of 2y under the cube root sign). In general, you need n factors to clear an nth root.*

*The process of eliminating a root in the denominator is called rationalizing the denominator.*

As another example, to simplify $\sqrt[4]{\frac{2x}{5}}$:

1. Apply the division property of radicals. $= \frac{\sqrt[4]{2x}}{\sqrt[4]{5}}$

2. Rationalize the denominator. $= \frac{\sqrt[4]{2x}}{\sqrt[4]{5}} \cdot \frac{\sqrt[4]{5^3}}{\sqrt[4]{5^3}}$

3. Perform the multiplication. $= \frac{\sqrt[4]{5^3 \cdot 2x}}{\sqrt[4]{5^4}}$

4. Simplify. $= \frac{\sqrt[4]{125 \cdot 2x}}{5}$

$$= \frac{\sqrt[4]{250x}}{5}$$

When there is a sum or difference involving roots in the denominator of a radical expression, you can often simplify the expression by multiplying the numerator and denominator by the conjugate of the denominator.

For example, to simplify $\frac{3}{\sqrt{x}+5}$:

1. Multiply the numerator and denominator by the conjugate of $(\sqrt{x}+5)$, $(\sqrt{x}-5)$. $= \frac{3}{\sqrt{x}+5} \cdot \frac{(\sqrt{x}-5)}{(\sqrt{x}-5)}$

$$= \frac{3(\sqrt{x}-5)}{(\sqrt{x}+5)(\sqrt{x}-5)}$$

2. Simplify. $= \frac{3 \cdot \sqrt{x} - 3 \cdot 5}{\left(\sqrt{x}\right)^2 - (5)^2}$

$$= \frac{3\sqrt{x}-15}{x-25}$$

## Sample Problems

1. Find: $5\sqrt{49} + 6\sqrt{3} + 15\sqrt{3} + 2\sqrt{48}$

   ☑ a. Factor the radicals into their prime factors. $= 5\sqrt{7 \cdot 7} + 6\sqrt{3} + 15\sqrt{3} + 2\sqrt{3 \cdot 4 \cdot 4}$

   ☑ b. Where possible, rewrite factors as perfect squares. $= 5\sqrt{7^2} + 6\sqrt{3} + 15\sqrt{3} + 2\sqrt{3 \cdot 4^2}$

   ☐ c. Take perfect squares out from under the radical signs. = ______________________

   ☐ d. Simplify. = ______________________

   ☐ e. Combine like terms. = ______________________

2. Find: $(\sqrt{x} + 5)(6\sqrt{x} - 8)$

   ☐ a. Find the sum of the products of the first terms, the outer terms, the inner terms, and the last terms. $= 6x - 8\sqrt{x} +$ _____ $-$ _____

   ☐ b. Combine like terms. = ______________________

3. Simplify: $\frac{5}{\sqrt[3]{x} \cdot \sqrt{y}}$

   ☑ a. Multiply the numerator and denominator by $(\sqrt[3]{x^2}) \cdot (\sqrt{y})$. $= \frac{5}{\sqrt[3]{x} \cdot \sqrt{y}} \cdot \frac{(\sqrt[3]{x^2}) \cdot (\sqrt{y})}{(\sqrt[3]{x^2}) \cdot (\sqrt{y})}$

   ☐ b. Simplify the radicals. = ______________________

*Answers to Sample Problems*

1.
c. $5 \cdot 7 + 6\sqrt{3} + 15\sqrt{3} + 2 \cdot 4\sqrt{3}$

d. $35 + 6\sqrt{3} + 15\sqrt{3} + 8\sqrt{3}$

e. $35 + 29\sqrt{3}$

2.
a. $30\sqrt{x}$, $40$

b. $6x + 22\sqrt{x} - 40$

3.
b. $\frac{5(\sqrt[3]{x^2}) \cdot (\sqrt{y})}{xy}$

# HOMEWORK

## Homework Problems

Circle the homework problems assigned to you by the computer, then complete them below.

### Explain

### Roots and Exponents

1. Rewrite using a radical, then evaluate: $8^{\frac{1}{3}}$

2. Evaluate: $\sqrt[3]{-216}$

3. Evaluate: $2^{\frac{1}{4}} \cdot 2^{\frac{2}{5}}$

4. Rewrite using a radical, then evaluate: $4^{\frac{3}{2}}$

5. Evaluate: $\sqrt[5]{1024}$

6. Simplify the expression below. Write your answer using only positive exponents.

$$\left(x^{\frac{1}{3}} \cdot y^{\frac{1}{2}}\right)^{-2}$$

7. Evaluate: $\sqrt[4]{-81}$

8. Simplify: $x^{\frac{4}{3}} \cdot x^{\frac{1}{3}}$

9. The number of cells of one type of bacteria doubles every 5 hours according to the formula $n_f = n_i \cdot 2^{\frac{t}{5}}$ where $n_f$ is the final number of cells, $n_i$ is the initial number of cells, and $t$ is the initial number of hours since the growth began. If a biologist starts with a single cell of the bacteria, how many cells will she have after 50 hours?

10. Alan invests $100 in a savings account. How much money would he have after a year if the interest rate for this account is 3% compounded every 4 months?

The amount $A$ in a savings account can be expressed as

$$A = P\left(1 + \frac{r}{n}\right)^{nt}$$

where $P$ is the amount of money initially invested, $t$ is the number of years the money has been invested, $r$ is the annual rate of interest, and $n$ is the number of times the interest is compounded each year.

11. Evaluate the expression below. Express your answer using only positive exponents.

$$\left(x^{\frac{5}{7}} \cdot x^{-\frac{3}{4}} \cdot y^{\frac{5}{4}}\right)^{-4}$$

12. Evaluate the expression below. Express your answer using only positive exponents.

$$\left(\sqrt[3]{-1331}\right)\left(x^{\frac{2}{9}}\right)^{-3}\left(\frac{1}{x^{-\frac{1}{3}}}\right)$$

### Simplifying Radicals

Simplify the expressions in problems (13)–(20). Assume $x$, $y$, and $z$ are positive numbers.

13. $\sqrt[3]{\frac{54}{250}}$

14. $\sqrt{\frac{49x}{4}}$

15. $\frac{\sqrt{5x^3}}{\sqrt{5x^2}}$

16. $\sqrt{\frac{-1048}{775}}$

17. $\sqrt[3]{-27x^6y^3}$

18. $\frac{\sqrt{x}}{\sqrt{y^4z^{12}}}$

19. $\sqrt[3]{16x^9y^4z^2}$

20. $\sqrt{\frac{242xy^{12}}{288y^2z^4}}$

21. One of the three unsolved problems of antiquity was to "double a cube"—that is, to construct a cube with twice the volume of a given cube. What would be the length of a side of a cube with twice the volume of 1 $m^3$? (Hint: The volume, $V$, of a cube with sides of length $L$ is $V = L \cdot L \cdot L = L^3$.)

22. The surface area, $A$, of a cube with sides of length $s$ is given by this formula:

$$s = \sqrt{\frac{A}{6}}$$

The volume, $V$, of the cube is:

$$V = s^3$$

What is the volume of a cube with a surface area of 48 $ft^2$?

Simplify the expressions in problems (23) and (24). Assume $x$, $y$, and $z$ are positive numbers.

23. $\sqrt{4x^5y^7z^2}$

24. $\dfrac{\sqrt[3]{-8x^6y^3z^3}}{\sqrt{36x^2y^6z^2}}$

## Operations on Radicals

25. Circle the like terms:

$7\sqrt[4]{360}$

$72\sqrt{360}$

$\dfrac{\sqrt[4]{360}}{72}$

$360\sqrt[4]{72}$

$-22\sqrt[3]{360}$

$-\dfrac{11}{4}\sqrt[4]{360}$

$\sqrt[4]{36}$

26. Simplify: $7\sqrt{125} + 2\sqrt{500} - \frac{3}{2}\sqrt{20} - 2\sqrt{10}$

27. Simplify: $5\sqrt{12}\left(6\sqrt{3} - 7\sqrt{27}\right)$

28. Circle the like terms:

| | |
|---|---|
| $79$ | $\sqrt{79}$ |
| $\sqrt{79^3}$ | $\sqrt[2]{79}$ |
| $\frac{1}{44}\sqrt{799}$ | $-\sqrt{79}$ |
| $2\sqrt{3} \cdot \sqrt{79}$ | $-\sqrt[2]{79^2}$ |

29. Simplify: $8\sqrt[3]{24} - \frac{\sqrt{16}}{2} + \frac{1}{4}\sqrt[3]{2} - \sqrt[3]{-3^3}$

30. Simplify: $\left(7\sqrt{2} - 8\sqrt{3}\right)\left(5\sqrt{2} + 6\sqrt{3}\right)$

31. Circle the like terms:

$\sqrt[3]{24}$

$2\sqrt[3]{3}$

$\sqrt[3]{8} \cdot \sqrt[3]{3}$

$\sqrt{(-3)^2}$

$\dfrac{\sqrt[3]{3}}{27}$

$3^{-3}$

$3^{\frac{1}{3}}$

32. Simplify: $\dfrac{1 - \sqrt{5}}{\sqrt{5} - 9}$

33. The period of a simple pendulum is given by the formula $t = 2\pi\sqrt{\frac{L}{32}}$ where $t$ is the period of the pendulum in seconds, and $L$ is the length of the pendulum in feet. What is the period of a 16 foot pendulum?

34. The Pythagorean Theorem, $a^2 + b^2 = c^2$, gives the relationship between the lengths of the two legs of a right triangle, $a$ and $b$, and the length of the hypotenuse of the triangle, $c$. If the lengths of the legs of a right triangle are $\sqrt{2}$ cm and $\sqrt{6}$ cm, how long is the hypotenuse?

35. Simplify: $\dfrac{\sqrt[3]{22}}{\sqrt[3]{77}}$

36. Simplify: $\dfrac{6\sqrt[3]{2} - 2\sqrt[3]{4}\left(2\sqrt[3]{32} - 2\sqrt[3]{4}\right)}{2\sqrt[3]{2} - \sqrt{2} \cdot \sqrt[3]{2}}$

# APPLY

## Practice Problems

Here are some additional practice problems for you to try.

### Roots and Exponents

1. Rewrite using a radical, then evaluate: $9^{\frac{5}{2}}$
2. Rewrite using a radical, then evaluate: $16^{\frac{3}{2}}$
3. Rewrite using a radical, then evaluate: $27^{\frac{2}{3}}$
4. Rewrite using a radical, then evaluate: $32^{\frac{4}{5}}$
5. Rewrite using a radical, then evaluate: $81^{\frac{3}{4}}$
6. Evaluate: $\sqrt[4]{625}$
7. Evaluate: $\sqrt[5]{7776}$
8. Evaluate: $\sqrt[5]{1024}$
9. Evaluate: $-\sqrt[4]{81}$
10. Evaluate: $\sqrt[5]{-32}$
11. Evaluate: $\sqrt[3]{-216}$
12. Rewrite using rational exponents: $\sqrt[4]{245^3}$
13. Rewrite using rational exponents: $\sqrt[5]{312^4}$
14. Rewrite using rational exponents: $\sqrt[3]{315^2}$
15. Rewrite using rational exponents: $\sqrt[7]{200^5}$
16. Rewrite using rational exponents: $\sqrt[8]{400^3}$
17. Find: $y^{\frac{2}{3}} \cdot y^{\frac{1}{4}}$
18. Find: $x^{\frac{1}{3}} \cdot z^{\frac{2}{5}}$
19. Find: $x^{\frac{1}{6}} \cdot x^{\frac{1}{5}}$
20. Find: $x^{\frac{1}{7}} \cdot x^{\frac{3}{7}} \cdot x^{\frac{2}{7}}$
21. Find: $x^{\frac{2}{9}} \cdot x^{\frac{5}{9}} \cdot x^{\frac{2}{9}}$
22. Find: $x^{\frac{3}{4}} \cdot x^{\frac{1}{2}} \cdot x^{\frac{3}{4}}$
23. Evaluate the expression below. Express your answer using only positive exponents.
$$\left(\frac{3a^{\frac{3}{4}}}{2b^2}\right)^{-4}$$
24. Evaluate the expression below. Express your answer using only positive exponents.
$$\left(\frac{x^{-\frac{4}{5}}}{2y}\right)^5$$
25. Evaluate the expression below. Express your answer using only positive exponents.
$$\left(\frac{4x^{-\frac{2}{3}}}{3y}\right)^3$$
26. Evaluate the expression below. Express your answer using only positive exponents.
$$\left(a^{\frac{3}{7}} \cdot b^{-\frac{2}{5}}\right)^3$$
27. Evaluate the expression below. Express your answer using only positive exponents.
$$\left(x^{-\frac{4}{9}} \cdot y^{\frac{6}{11}}\right)^2$$
28. Evaluate the expression below. Express your answer using only positive exponents.
$$\left(x^{-\frac{2}{3}} \cdot y^{\frac{3}{5}} \cdot z^{-\frac{4}{7}}\right)^3$$

### Simplifying Radicals

29. Simplify: $\sqrt{\frac{121}{64}}$
30. Simplify: $\sqrt{\frac{289}{361}}$
31. Simplify: $\sqrt{\frac{169}{576}}$
32. Simplify: $\sqrt[3]{\frac{27}{8}}$
33. Simplify: $\sqrt[3]{\frac{-64}{125}}$

34. Simplify: $\sqrt[3]{\frac{-343}{27}}$

35. Simplify: $\sqrt[5]{\frac{-32}{243}}$

36. Simplify: $\sqrt[4]{\frac{625}{1296}}$

37. Simplify: $\sqrt[6]{\frac{729}{64}}$

38. Calculate: $\sqrt{(-35^2)}$

39. Calculate: $\sqrt{(-56)^2}$

40. Calculate: $\sqrt[3]{(13^3)}$

41. Calculate: $\sqrt[5]{(-47^5)}$

42. Calculate: $\sqrt[3]{(-29)^3}$

43. Which of the radical expressions below is in simplified form?

$\frac{\sqrt{81}}{x}$ $\sqrt{\frac{25}{49}}$ $\frac{6}{\sqrt{30}}$ $\frac{\sqrt[4]{7}}{x}$

44. Which of the radical expressions below is in simplified form?

$\frac{\sqrt[3]{5}}{x}$ $\frac{4}{\sqrt{20}}$ $\sqrt{\frac{16}{9}}$ $\frac{\sqrt{49}}{x}$

45. Simplify: $\sqrt{36a^2b^6}$

46. Simplify: $\sqrt{100m^6n^4}$

47. Simplify: $\sqrt{64x^4y^6z^{10}}$

48. Simplify: $\sqrt{54a^3b^8}$

49. Simplify: $\sqrt{108m^5n^9}$

50. Simplify: $\sqrt{72x^4y^7}$

51. Simplify: $\sqrt[3]{192a^3b^5c^9}$

52. Simplify: $\sqrt[3]{-250x^4y^6z^8}$

53. Simplify: $\sqrt[5]{160m^2n^7p^{12}}$

54. Simplify: $\frac{\sqrt{49a^3b^8}}{\sqrt{7ab^7}}$

55. Simplify: $\frac{\sqrt[3]{64m^7n^5}}{\sqrt[3]{2mn^3}}$

56. Simplify: $\frac{\sqrt[4]{81x^9y^6}}{\sqrt[4]{3xy^4}}$

## Operations on Radicals

57. Circle the like terms:

$\frac{\sqrt{-5}}{4}$

$\sqrt{5}$

$\frac{1}{3}\sqrt{5}$

$-9\sqrt{50}$

$\frac{7\sqrt{5}}{8}$

$\frac{6\sqrt[5]{50}}{13}$

$-\sqrt[5]{2}$

58. Circle the like terms:

$\frac{5}{2}\sqrt{3}$

$-\sqrt[3]{2}$

$\frac{\sqrt{-3}}{3}$

$-6\sqrt{30}$

$\frac{4\sqrt[3]{30}}{15}$

$\sqrt{3}$

$\frac{6\sqrt{3}}{5}$

59. Simplify: $7\sqrt{5} + \sqrt{20} - 3\sqrt{80}$

60. Simplify: $10\sqrt{2} - \sqrt{128} + 3\sqrt{32}$

61. Simplify: $8\sqrt{3} + \sqrt{12} - 4\sqrt{27}$

62. Simplify: $\sqrt{40} + 3\sqrt{10} - \sqrt{18}$

63. Simplify: $4\sqrt{50} - 5\sqrt{27} + 2\sqrt{75}$

64. Simplify: $\sqrt{20} - 2\sqrt{18} + \sqrt{8}$

65. Simplify: $\sqrt[3]{250x^2} + 3\sqrt[3]{16x^5} - 3\sqrt[3]{432x^2}$

66. Simplify: $5\sqrt[4]{32y} + \sqrt[4]{162y^5} - \sqrt[4]{1250y}$

67. Simplify: $\sqrt[3]{128x} + 2\sqrt[3]{16x^4} - \sqrt[3]{54x}$

68. Simplify: $5\sqrt{6}\left(3\sqrt{8} - 9\sqrt{21}\right)$

69. Simplify: $2\sqrt[3]{9}\left(5\sqrt[3]{3} - 7\sqrt[3]{5}\right)$

70. Simplify: $3\sqrt[4]{8}\left(4\sqrt[2]{2} + 6\sqrt[4]{3}\right)$

71. Simplify: $3\sqrt{2y}\left(7\sqrt{10y} + 4\sqrt{3}\right)$

72. Simplify: $6\sqrt{3z}\left(2\sqrt{6z} - 3\sqrt{5}\right)$

73. Simplify: $2\sqrt[3]{4z}\left(5\sqrt[3]{2z^2} - 7\sqrt[3]{11z}\right)$

74. Simplify: $\left(\sqrt{5} + \sqrt{3}\right)\left(\sqrt{6} + \sqrt{11}\right)$

75. Simplify: $\left(\sqrt{5} - \sqrt{10}\right)\left(\sqrt{2} - \sqrt{15}\right)$

76. Simplify: $\left(\sqrt{6} + \sqrt{5}\right)\left(\sqrt{3} - \sqrt{10}\right)$

77. Simplify: $\left(3\sqrt{5z} + \sqrt{6}\right)\left(3\sqrt{5z} - \sqrt{6}\right)$

78. Simplify: $\left(2\sqrt{3y} + \sqrt{7}\right)\left(2\sqrt{3y} - \sqrt{7}\right)$

79. Simplify: $\left(5\sqrt{2y} - \sqrt{3x}\right)\left(5\sqrt{2y} + \sqrt{3x}\right)$

80. Simplify: $\frac{3\sqrt{y}}{\sqrt{6y}}$

81. Simplify: $\frac{2\sqrt{x}}{\sqrt{2x}}$

82. Simplify: $\frac{3\sqrt{5}}{x + \sqrt{5}}$

83. Simplify: $\frac{5\sqrt{2}}{x - \sqrt{2}}$

84. Simplify: $\frac{x - \sqrt{3}}{x + \sqrt{3}}$

# EVALUATE

## Practice Test

Take this practice test to be sure that you are prepared for the final quiz in Evaluate.

Assume that *x*, *y*, and *z* are positive numbers.

1. Simplify: $\sqrt[5]{x} \cdot \sqrt{x}$

2. Rewrite the expression using rational exponents.

   $\sqrt[5]{243^3}$

3. Circle the real number(s) in the list below:

   $\sqrt{-100}$

   $\sqrt[3]{-125}$

   $\sqrt[4]{-16}$

   $\sqrt[6]{-729}$

4. Simplify: $\left(\dfrac{8y^{-\frac{1}{2}}}{7^{\frac{3}{2}}x}\right)^2$

5. Simplify: $\sqrt{\dfrac{169}{225}}$

6. Calculate: $\sqrt{(-29)^2}$

7. Which of the radical expressions below is simplified?

   $\sqrt{\dfrac{3}{16}}$

   $\dfrac{xy}{\sqrt{8}}$

   $\dfrac{\sqrt{25}}{y}$

   $\dfrac{\sqrt[3]{3}}{2}$

8. Simplify: $\sqrt{\dfrac{81x^2y^2}{121z}}$

9. Simplify: $6\sqrt{5x} + 3\sqrt{125x} - 3$

10. Find: $(3\sqrt{5} - 8)(3\sqrt{5} + 8)$

11. Find: $(3\sqrt{2} + 3)(2\sqrt{2} - 6)$

12. Find: $\dfrac{\sqrt{y}}{\sqrt[3]{y}}$

# ANSWERS

## Homework Problems

**1.** $\sqrt[3]{8} = 2$ **3.** $2^{\frac{13}{20}}$ **5.** 4 **7.** not a real number

**9.** 1024 cells **11.** $\frac{x^{\frac{1}{7}}}{y^5}$ **13.** $\frac{3}{5}$ **15.** $\sqrt{x}$ **17.** $-3x^2y$

**19.** $2x^3y\sqrt[3]{2yz^2}$ **21.** $\sqrt[3]{2}m$ **23.** $2x^2y^3z\sqrt{xy}$

**25.** $7\sqrt[4]{360}$, $\frac{\sqrt[4]{360}}{72}$, $\frac{-11}{4}\sqrt[4]{360}$

**27.** $-450$ **29.** $16\sqrt[3]{3} + \frac{1}{4}\sqrt[3]{2} + 1$

**31.** $\sqrt[3]{24}, 2\sqrt[3]{3}, \sqrt[3]{8}\cdot\sqrt[3]{3}, \frac{\sqrt[3]{3}}{27}, 3^{\frac{1}{3}}$ **33.** $\sqrt{2}\pi$

**35.** $\frac{\sqrt[3]{98}}{7}$

## Practice Problems

**1.** $\sqrt{9^5}$ or $\left(\sqrt{9}\right)^5$; 243 **3.** $\sqrt[3]{27^2}$ or $\left(\sqrt[3]{27}\right)^2$; 9

**5.** $\sqrt[4]{81^3} = \left(\sqrt[4]{81}\right)^3 = 27$ **7.** 6 **9.** $-3$ **11.** $-6$

**13.** $312^{\frac{4}{5}}$ **15.** $200^{\frac{5}{7}}$ **17.** $y^{\frac{11}{12}}$ **19.** $x^{\frac{11}{30}}$ **21.** $x$

**23.** $\frac{16b^8}{81a^3}$ **25.** $\frac{64}{27x^2y^3}$ **27.** $\frac{y^{\frac{12}{11}}}{x^{\frac{8}{9}}}$

**29.** $\frac{11}{8}$ **31.** $\frac{13}{24}$ **33.** $-\frac{4}{5}$ **35.** $-\frac{2}{3}$ **37.** $\frac{3}{2}$

**39.** 56 **41.** $-47$ **43.** $\frac{\sqrt[4]{7}}{x}$ **45.** $6ab^3$ **47.** $8x^2y^3z^5$

**49.** $6m^2n^4\sqrt{3mn}$ **51.** $4abc^3\sqrt[3]{3b^2}$ **53.** $2np^2\sqrt[5]{5m^2n^2p^2}$

**55.** $2m^2\sqrt[3]{4n^2}$ **57.** $\sqrt{5}, \frac{1}{3}\sqrt{5}, \frac{7\sqrt{5}}{8}$ **59.** $-3\sqrt{5}$

**61.** $-2\sqrt{3}$ **63.** $20\sqrt{2} - 5\sqrt{3}$

**65.** $6x\sqrt[3]{2x^2} - 13\sqrt[3]{2x^2}$ or $(6x - 13)\sqrt[3]{2x^2}$

**67.** $4x\sqrt[3]{2x} - 11\sqrt[3]{2x}$ or $(4x - 11)\sqrt[3]{2x}$ **69.** $30 - 14\sqrt[3]{45}$

**71.** $42y\sqrt{5} + 12\sqrt{6y}$ **73.** $20z - 14\sqrt[3]{44z^2}$

**75.** $\sqrt{10} - 5\sqrt{3} - 2\sqrt{5} + 5\sqrt{6}$ **77.** $45z - 6$

**79.** $50y - 3x$ **81.** $\frac{\sqrt{2x}}{x}$ **83.** $\frac{5x\sqrt{2} + 10}{x^2 - 2}$

## Practice Test

**1.** $x^{\frac{7}{10}}$ **2.** $243^{\frac{3}{5}}$ **3.** $\sqrt[3]{-125}$ **4.** $\frac{64}{343x^2y}$

**5.** $\frac{13}{15}$ **6.** 29 **7.** $\frac{\sqrt[3]{3}}{2}$ **8.** $\frac{9xy\sqrt{z}}{11z}$

**9.** $21\sqrt{5x} - 3$ **10.** $-19$ **11.** $-6 - 12\sqrt{2}$ **12.** $y^{\frac{1}{6}}$

# TOPIC 9 CUMULATIVE ACTIVITIES

## CUMULATIVE REVIEW PROBLEMS

These problems combine all of the material you have covered so far in this course. You may want to test your understanding of this material before you move on to the next topic. Or you may wish to do these problems to review for a test.

1. Find: $(x^2 + 12x)(x + 3y^2 + 1)$
2. Solve for $x$: $\frac{2x + 5}{2 - x} = 3$
3. Solve for $y$: $\frac{1}{y} - \frac{2}{3} = \frac{y}{3}$
4. Find:
   a. $(125)^{\frac{1}{3}}(16)^{\frac{3}{4}}$
   b. $(x^3y)^{\frac{1}{3}}$
   c. $\frac{a^{\frac{1}{2}}b^2a^{\frac{1}{3}}}{b^{-2}}$
5. Last year Scott earned 5% in interest on his savings account and 13% in interest on his money market account. If he had \$14,125 in the bank and earned a total of \$1706.25 in interest, how much did he have in each account?
6. Graph the line that passes through the point (0, –3) with slope 2.
7. For what values is the rational expression $\frac{x^3 - 3x + 29}{x^2 + 13x + 36}$ undefined?
8. Solve $-10 < 9x - 7 < 11$ for $x$.
9. Solve this system of equations:

$$y = -\frac{2}{7}x + 3$$
$$14y + 4x = 14$$

10. Factor: $2ab + 14a + 5b + 35$
11. Angela and Casey were asked to clean their classroom. Working alone, Angela could clean the room in 20 minutes. It would take Casey 25 minutes to clean the room by herself. How long would it take them to clean the room together?
12. Simplify: $\left(\frac{27}{x}\right)^{-\frac{1}{3}}\left(\sqrt{x}\right)^{\frac{4}{3}}$
13. Find the equation of the line that passes through the point (–7, 3) and has slope $\frac{5}{3}$. Write your answer in point-slope form, in slope-intercept form, and in standard form.
14. Simplify this expression: $2r^2s + 3t + 4s^2 - 5r^2s - 6s^2 + 7t$
15. Find:
    $(3a^2b^2 + 2a^2b - 7ab + a) - (a^2b^2 - 12ab + 2a^2b + b)$
16. Simplify: $\frac{2 + \sqrt{2}}{\sqrt{2} + \sqrt{6}}$
17. Find:
    a. $\frac{3^0 \cdot 10^2}{2^3}$
    b. $(-3a^2)^3$
    c. $[(x^3y^2)^2z]^4$
18. Graph the inequality $2y - 10x \leq 32$.
19. Solve for $x$: $3(x + 2) - x = 3x - 8$
20. Graph the line $y - \frac{1}{2} = \frac{1}{6}(x + 2)$.
21. Factor: $-16y^2 + 24y - 9$
22. Rewrite using radicals, then simplify: $\frac{\left(24^{\frac{1}{3}} + 100^{\frac{1}{2}}\right)}{16^{\frac{1}{4}}}$

23. Circle the true statements.

$22 + 32 = 52$

$|3 - 4| = |3| - |4|$

The GCF of 52 and 100 is 4.

$\frac{9}{25} = \frac{3}{5}$

The LCM of 30 and 36 is 180.

24. Find the slope of the line that is perpendicular to the line that passes through the points (8, 2) and (–4, 9).

25. Factor: $x^2 + 3x - 130$

26. In a bin, the ratio of red apples to green apples is 10 to 3. If there is a total of 15 green apples, how many red ones are there?

27. Find the slope and $y$-intercept of this line: $4y + 3x = -18$

28. Graph the system of linear inequalities below to find its solution.

$3y - 5x < 3$

$5y + 3x > -10$

29. Simplify: $\sqrt{\frac{72x^3y^2}{(2y^2)^2}}$

30. Find the slope of the line through the points $\left(\frac{3}{4}, 7\right)$ and $\left(\frac{1}{4}, -\frac{1}{2}\right)$.

31. Simplify: $\frac{8}{3 + \sqrt{11}}$

32. Graph the line $2y + 1 = 1 - 3x$.

33. Find: $(2x^3 + 21x^2 - 27x + 8) \div (2x - 1)$

34. Evaluate the expression $3a^2 + ab - b^2$ when $a = 2$ and $b = 8$.

35. Factor: $7x^2y^2 + 14xy^2 + 7y^2$

36. Factor: $5x^2 - 80$

37. Solve for $y$: $-10 < 5 - 3y \leq 2$

38. Rewrite using only positive exponents: $\frac{a^3b^{-2}}{(c^{-1})^{-2}}$

39. Find:

a. $\sqrt{20} + \sqrt{80}$

b. $\sqrt{6}\left(\sqrt{6x^2} + \sqrt{3x^2}\right)$

c. $\left(a + \sqrt{b}\right)\left(a - \sqrt{b}\right)$

40. Factor: $a^2 + 6a + 6b + ab$

41. A juggler has 10 more balls than juggling pins. If the number of balls is 1 more than twice the number of pins, how many pins and balls are there?

42. Find: $\frac{x^2 - 9}{x^2 + 3x} \cdot \frac{x^2 - 7x}{x^2 - 10x + 21}$

43. Simplify: $\frac{x - y}{\sqrt{x} - \sqrt{y}}$

44. Solve for $y$: $3\left[7y + 5(1 - 2y)\right] = -27$

45. Factor: $8x^3 - 1$

46. Solve for $x$: $\frac{1}{2}(x + 7) = 12$

47. Solve for $a$: $\frac{a}{a - 5} + 1 = \frac{a - 3}{a - 5}$

48. Evaluate the expression $a^3b + 3 - ab^3 + 2ab$ when $a = -2$ and $b = 4$.

49. Simplify: $\left(\frac{2y^3}{3x^4}\right)^2$

50. Factor: $2x^2 - 40x + 198$

# ANSWERS

## Cumulative Review Problems

**1.** $x^3 + 3x^2y^2 + 36xy^2 + 13x^2 + 12x$ **3.** $y = 1$ or $y = -3$

**5.** Loan = $625 **7.** $x = -4$ or $x = -9$ **9.** no solution

**11.** $11\frac{1}{9}$ minutes

**13.** point-slope form: $-3 = \frac{5}{3}(x + 7)$

slope intercept form: $y = \frac{5}{3}x + \frac{44}{3}$

standard form: $3y - 5x = 44$

**15.** $2a^2b^2 + 5ab + a - b$

**17a.** $\frac{25}{2}$ **b.** $-27a^6$ **c.** $x^{24}y^{16}z^4$ **19.** $x = 14$

**21.** $(-4y - 3)(4y + 3)$ or $(-4y + 3)(4y - 3)$ or $-(4y - 3)^2$

**23.** True: The GCF of 52 and 100 is 4. The LCM of 30 and 36 is 180.

**25.** $(x + 13)(x - 10)$ **27.** slope $= -\frac{3}{4}$, $y$-intercept $= (0, -\frac{9}{2})$

**29.** $\frac{3x\sqrt{2x}}{y}$ **31.** $\frac{24 - 8\sqrt{11}}{-2}$ or $-12 + 4\sqrt{11}$

**33.** $x^2 + 11x - 8$ **35.** $7y^2(x + 1)^2$

**37.** $1 \leq y < 5$ **39a.** $6\sqrt{5}$ **b.** $6x + 3x\sqrt{2}$ **c.** $a^2 - b$

**41.** 9 pins, 19 balls **43.** $\sqrt{x} + \sqrt{y}$

**45.** $(2x - 1)(4x^2 + 2x + 1)$ **47.** $a = 2$ **49.** $\frac{4y^6}{9x^8}$

# TOPIC 9 INDEX